GREENBACK SERIES: NUMBER NINE

The Spread of BRITISH PRINTING 1557 to 1695

'DO NOT THINK THAT YOU CAN BLIND THE CROWS'

by

William K. Sessions, MA (Econ), FIOP

ISBN 1 85072 036 3

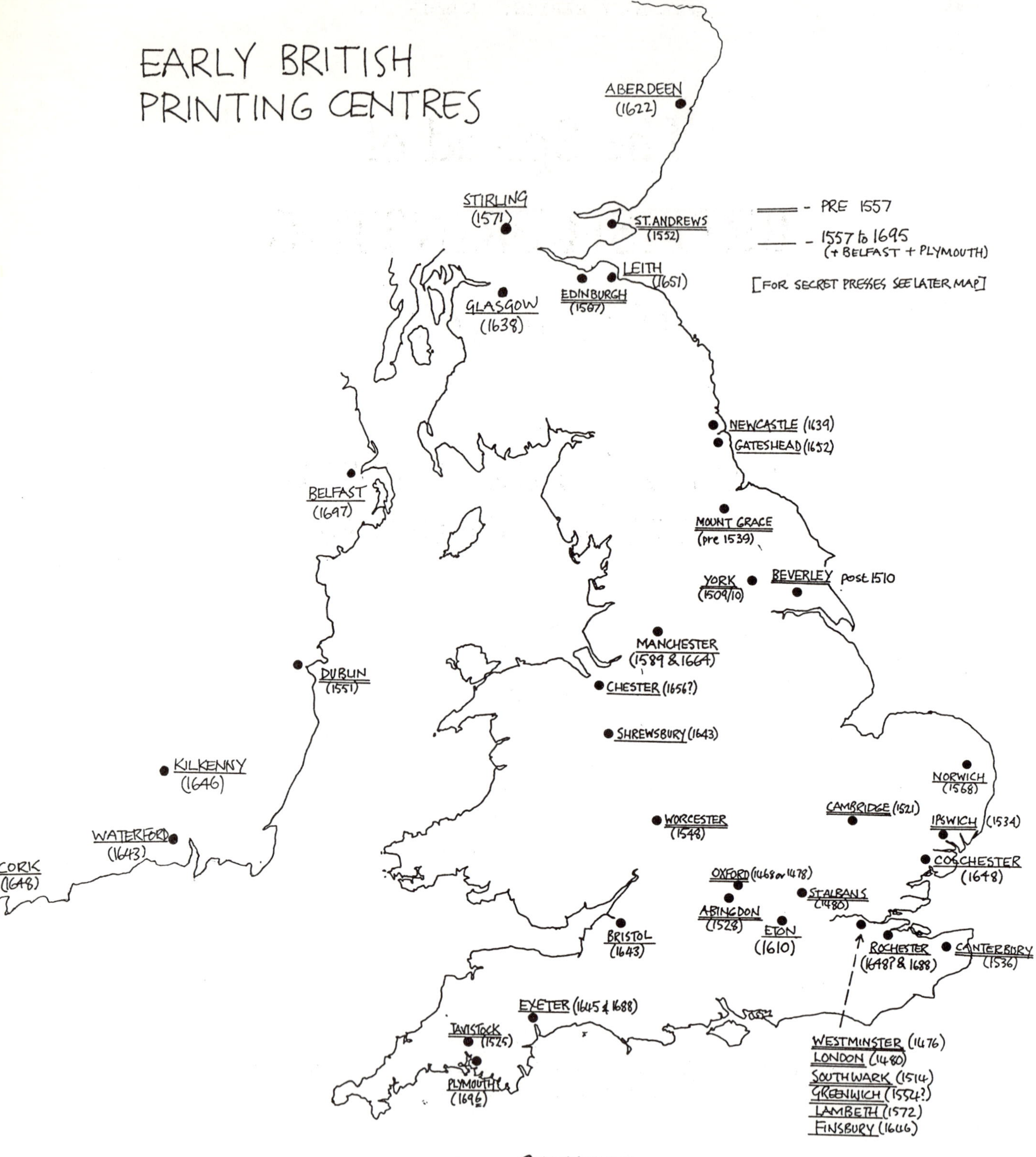
EARLY BRITISH
PRINTING CENTRES
ABERDEEN (1622)
STIRLING (1571)
ST. ANDREWS (1552)
LEITH (1651)
EDINBURGH (1507)
GLASGOW (1638)
- PRE 1557
- 1557 to 1695 (+ BELFAST + PLYMOUTH)
[FOR SECRET PRESSES SEE LATER MAP]
NEWCASTLE (1639)
GATESHEAD (1652)
BELFAST (1697)
MOUNT GRACE (pre 1539)
YORK (1509/10)
BEVERLEY post 1510
MANCHESTER (1589 & 1664)
DUBLIN (1551)
CHESTER (1656?)
SHREWSBURY (1643)
KILKENNY (1646)
NORWICH (1568)
CAMBRIDGE (1521)
IPSWICH (1534)
WORCESTER (1548)
WATERFORD (1643)
CORK (1648)
COLCHESTER (1648)
OXFORD (1468 or 1478)
ST. ALBANS (1480)
ABINGDON (1528)
ETON (1610)
BRISTOL (1643)
ROCHESTER (1648? & 1688)
CANTERBURY (1536)
EXETER (1645 & 1688)
TAVISTOCK (1525)
PLYMOUTH (1696)
WESTMINSTER (1476)
LONDON (1480)
SOUTHWARK (1514)
GREENWICH (1554?)
LAMBETH (1572)
FINSBURY (1646)

C O N T E N T S

ILLUSTRATIONS

INTRODUCTION

In the first of my (now) nine green-back studies, PRINTERS DOZEN (see list on the back cover), I summarised the printing activities in about a dozen British locations outside Westminster and London, subsequent to Caxton's British printing arrival in 1476 until the granting of the monopolistic printing charter to the London Stationers Company on 4th May 1557.

This present volume summarises all known British printing locations from 1557, for nearly 140 years until the year 1695, which marked the lapsing of the final Licensing Act, after which date English provincial printing of books, newsheets and ephemera "took off in all directions".

The Fourth of May continues to be marked by an annual dinner of the currently named Worshipful Company of Stationers and Newspaper Makers (a splendid event which I attended in 1987 as a liveryman since 1977 of this historic London Company).

For it was on the 4th May 1557 that the privy seal was set to the charter of incorporation of the Stationers Company. As Cyprian Blagden writes in his History of the Stationers' Company (page 21):- The preamble to the charter declares that the King and Queen (Philip and Mary Tudor):-

> wishing to provide a suitable remedy against the seditious and heretical books which were daily printed and published, gave certain privileges to their beloved and faithful lieges, the ninety-seven Stationers, in addition to the normal rights of a company. It was laid down, firstly, that no one in the realm should exercise the art of printing, either himself or through an agent, unless he were a freeman of the Stationers' Company of London or unless he had royal permission to do so; and secondly that the Master and Wardens of the Company were to have the right to search the houses and business premises of all printers, bookbinders and booksellers in the kingdom for any printed matter, to seize (and treat as they thought fit) anything printed contrary to any statute or proclamation, and to imprison anyone who printed without the proper qualification or resisted their search; such offenders were to remain in gaol for three months without trial and be fined £5, half of which was to go to the Crown and half to the Company.

Blagden adds that this nation-wide right of search was a privilege which had earlier been granted for very different reasons to the Goldsmiths and the Pewterers Companies.

Blagden also points out (p 40) that, of the ninety-seven named in the 1557 charter, only thirty-three were master printers. Journeymen printers could become "free of the Company" by completing an apprenticeship (usually of seven years) with a master-printer liveryman of the Company. Others could obtain entry (as now) through their father's membership, i.e. by patrimony. Members of other London Guilds could apply for Stationers' membership - 'by translation'. Blagden also adds that of these original ninety-seven charter signatories 'less than a score' were Booksellers. All in all this 1557 charter gave very considerable, exclusive powers to a total of just thirty-three master-printers practising their skilled 'black art' in the London area.

The motive of this Mary Tudor charter was clearly to restrict and to keep control of the dangerous vehicle of printing. The history of England would have been very different if Catholic Mary Tudor (daughter of Katherine of Aragon and wife of Philip II of Spain) had not died of a tumour at the early age of 43, in the following year, 1558.

However the half-sister who succeeded her, Elizabeth I confirmed this Stationers Company Charter on 10th November 1559, still with the objective of keeping a tight control on this feared printing instrument, but for the exactly opposite reason of enlisting its aid for the Protestant - not the Catholic - cause.

As will be seen in later chapters, earlier charters also permitted presses at each of the only two English universities of Oxford and Cambridge; but apart from these and those under the Stationers Company jurisdiction, no printing presses were permitted anywhere in England for about 140 years, or more precisely until the lapsing of the final Licensing Act in 1695, save for the addition of York in 1649. In these following pages I will describe and illustrate how many or how few were the exceptions.

Towards the end of this work, I will indicate graphically the dominant relative scale of London printing during this near century-and-a-half. I will also describe and attempt a numerical assessment of the uninterrupted printing achievements at both Cambridge and Oxford. For comparative purposes I will also summarise the somewhat spasmodic continuation of printing in Protestant Dublin "within the pale"; and I will also attempt to describe and indicate the scale of the uninterrupted continuation of printing in Scotland, which was a separate country until the Act of Union of 1707, more than a century after the union of the crowns by James the sixth of Scotland and first of England from 1603.

In my earlier 'green-back' study on Canterbury's first printer (see back-cover) I have already described and shown facsimiles (courtesy of both the Bodleian and Archbishop Marsh's Library in Dublin) of what is believed to be the only piece of printing produced during Mary Tudor's five-year reign in England outside London, Oxford and Cambridge. (See my Canterbury 'green-back' pages 18, 19, 49 and 50). However these Articles to be enquiyred in thordinary visitacion ....." were dated 1556 and so anticipate the Stationers Charter by one year.

In other of my 'green-back' series I have already described specific areas of printing activities in the English provinces during this 1557 to 1695 period, namely:-

(a) The First Printers in Norwich from 1567: Anthony de Solempne, Albert Christiaensz, and [perhaps] Joannes Paetz. The reason for this short-lived burst of Norwich printing, mostly in the Dutch language, was that a few printers and booksellers were among the waves of Protestant religious refugees who fled to Norwich across the North Sea from the Low Countries. This was just before the arrival in Brussels in August 1567 of the feared Spanish military leader, the Duke of Alva, with his 1,200 cavalry and his four regiments of catholic mercenaries. This printing by religious refugees is seemingly specific to Norwich without affecting the trend of printing in other areas of England. Yet it raises many intriguing queries about lack of Stationers' Company sanction and seemingly delayed royal approval.

(b) Then for the following century there are my two 'green-backs' describing and illustrating the printing activities of King Charles the First's travelling printers - initially in Newcastle upon Tyne in 1639 - and then three years later a separate series of journeys to York in 1642, to Shrewsbury in 1642-1643, to Bristol 1643-1645, and to Exeter in 1645-1646. My works include virtually all the facsimile imprints from those epic journeys, just prior to and during the Civil War, and I notably illustrate (courtesy of the National Library of Scotland) the travelling printers' own contemporary and detailed catalogue of works produced by them at York and at Shrewsbury. This has made possible, on a secure base, an estimate of the dramatically different ratios of surviving titles, in the months before the fighting began as compared with the first winter of the Civil War. The second work also disseminates (for the first time outside the confines of Exeter libraries) details of the interesting fate in 1646 of these travelling presses.

(c) Another of my 'green-backs':- <u>Bulkley & Broad: White & Wayt</u> is a 'block-buster' of 286 A4 pages including no less than 185 contemporary title-page facsimiles. This work firstly illustrates and describes the brave and dramatic activities of the ardent royalist Stephen Bulkley who fled with his printing apparatus in 1642 from London to royalist York; who later in 1646 arrived in Scottish-held Newcastle upon Tyne to aid the captive King, again mysteriously with his printing apparatus intact (despite the 1644 battle of Marston Moor and capitulation of York). Stephen Bulkley perforce stayed on in Newcastle when the King was controversially handed over to the English Parliamentarians and taken south in February 1646/7. From 1653-1659 Bulkley moved across the Tyne Bridge to the south bank of Gateshead, where he continued printing between 1653 and 1659, within the ecclesiastical jurisdiction of Durham. Between 1659 to 1662 he was again printing on the north bank in Newcastle; and finally he completed his varied and active printing career in post-Restoration York from 1663 to 1677. All in all an epic 35 year printing saga spanning London-York-Newcastle-Gateshead-Newcastle-York, with certainly one and possibly two 'moonlight-flits' with his printing apparatus always ready to risk all in the royalist cause.

This same Bulkley and Broad 'green-back' continues to describe and illustrate the 1644, post-Marston-Moor arrival in York of the Parliamentarian printer Thomas Broad.

As to Broad's pre-York antecedents I express grateful appreciation to printing-historian Sheila Lambert of 30 Millington Road, Cambridge for removing a mote or beam from the present author's eye [Not, please plank or sawdust - Matthew 7²]. She has most helpfully pointed out the following entry number 207 on page 16 of D. F. McKenzie's standard work on my shelves: Stationers' Company Apprentices 1605-1640. This states that Thomas Broad was apprenticed to the London Stationer Miles Flesher; that Broad's father was also called Thomas, a Clerk of Rendcombe in Gloucestershire; that young Thomas was bound apprentice on 7-5-1627, and obtained his freedom on 3-6-1634 after an apprenticeship period of seven years. [If this took place at the age of 21 or 22, Thomas Broad's date of birth would have therefore been 1612 or 1613]. Sheila Lambert helpfully also informs me as follows:

> In 1638 [Thomas Broad] signed the second journeymen's petition: Tanner MS 67 f. 194.
>
> In 1641 the poll tax lists show Thomas Broads [sic], stationer, lodging in the house of Wm Blackborne (profession not stated) in Ellis Ct, Old Bailey (that is Plomer's 'Eliots' Court).
>
> The poll tax lists also show a Jo. Bulkeley*, printer, lodging with Thos Hatt, attorney of Common Pleas, in St Botolph Ward.
>
> Neither of them was married at this time, since these lists show if 'et ux': T.C.Dale, The poll tax for London in 1641 (Society of Genealogists 1935, typescript) There is a copy in BL: Ac.5962 b/46. The original lists are in PRO: E.179/252/1-13.

* Stephen Bulkley's father was Joseph Buckley (but a stationer of Canterbury) - see my Bulkley & Broad....green-back p 1.

In York Thomas Broad printed for the Parliamentarian cause from the autumn of 1644, with his last proven extant York imprints being, as shown, in 1656. Whether there was an actual cessation of smaller pieces of York printing during the years 1657 and 1660 seems on first principles unlikely for (a) printing had been authorised in York by Parliamentarian statute in 1649; and (b) Thomas Broad was, as just described, a member of the London Stationers Company through apprenticeship-completion.

Although no death details have so far been traced for Thomas Broad (possibly in puritan or presbyterian records) it is presumed that he died in York circa 1660, at the conjectural age of only 47 or 48, because his widow Alice Broad(e) began a series of active York imprints for a further 19 years from 1661 to 1680. In that year John White came from London to marry (virtually certainly) Thomas and Alice's daughter Hannah Broad. He had been bookselling at the sign of The Three Bibles in The Minories in London (see my <u>Bulkley and Broad</u> p 241) and in 1680 he shared one imprint with his presumed mother-in-law "at the Sign of the Lyon and the Lamb" in Stonegate, York. (see p 242).

In turn John White continued York's printing narrative past our chosen 1695 ending-date, for he died in 1716. His extant titles show a broadening range of subject-matter, and he is perhaps chiefly famous for daring to print in York a sadly non-extant manifesto in support of the future William III's 1688 landing in the south-west. For this White was temporarily incarcerated in Hull Castle, but was subsequently awarded the office of "Their Majesties Printer for the City of York and the Five Northern Counties" (see title page facsimiles in my <u>Bulkley and Broad</u>.... pages 258-261).

Earlier, as already indicated the much travelled royalist printer Stephen Bulkley, late in 1662 or early in our 1663, had returned to York, doubtless in consequence of Charles II's Act of 1662 which added York to the cities of London, Oxford and Cambridge as the only places in England where printing was permitted. (This confirmed afresh the similar York printing authorisation by Parliamentarian statute in 1649). Thus from 1663 Stephen Bulkley was printing in York in competition with Thomas Broad's widow Alice. Following Stephen Bulkley's death in York in January 1679/1680, his youngest son John continued printing there, with but a very few extant imprints in 1683, 1684 and 1689; though from 1680 John White was busily printing in the City.

So all in all, York's printing narrative can be regarded as continuous from 1642 to the present day, and with official Parliamentarian statute from 1649 and with Charles II's Statute from 1662. From the standpoint of the Stationers Company Charter it should be noted that Stephen Bulkley, Thomas Broad and John White were all London Stationers.

....

So with all of these widely scattered provincial printing activities already described in my series of eight "green-back" studies, what other printing locations remain to be described in this present volume for the period 1557 to 1695?

There will now follow a chapter on Printing at Eton in Greek type from 1610.

Next I will bravely/foolhardily attempt summaries of the shadowy activities of a series of secret presses in a variety of locations in provincial England, plus one in a Welsh cave. There will follow a section on the four-in-one career of Robert Waldegrave - namely London: on-the-run: Scottish King's Printer: and back to London with James sixth and first in 1603.

Following this, I insert outline narratives of steadily continuing printing activity at Cambridge and at Oxford; also for comparative purposes in Ireland and in Scotland. I also touch briefly on some London area locations, and summarise two 1688 printing flurries at Chester and at Exeter.

And finally (helped by Ian Maxted of Guildhall Library and now Exeter) we will include charts showing numbers of extant, listed, titles year by year for several locations in context with a year by year chart showing the extant book output dominance by London printers during this designated period 1557 to 1695.

II

PRINTING AT ETON: 1610 - 1613

Eton College in Windsor, Berkshire was, like King's College, Cambridge, founded by King Henry VI (1422-1461). In the early 17th Century Eton College housed the famous private press of Sir Henry Savile, who was both Provost of Eton and also Warden of Merton College in Oxford.

Archdeacon Henry Cotton in his *Typographical Gazetteer Attempted* (1825) quotes from "Bagford's manuscript papers, at present contained in the Harleian Library as follows:-

> Sir Henry Saville, meditating an edition of St. Chrysostom, prepared a fount of curious Greek letters, which in those days were vulgarly called *the silver letter*, not being cast in silver, but for the beauty of the letter so called. He then made provision of presses and other materials for the undertaking, and resolved to print in Eton college, and there set up his presses, and about the year 1607 he printed some small pieces in Greek before he went in hand with the great work of St Chrysostom; and John Norton was the printer. These Greek letters came afterwards into the hands of one [William] Turner, a printer at Oxford.

According to E A Clough's *Short-Title Catalogue Arranged Geographically* ...... (published in 1969 and therefore based on the STC First Edition), the year of 1607 may indeed have been the date of Eton's first printing, with the Greek title translated as *Dionysius, Periegetes*. Courtesy of Cambridge University Library, I enclose title page of this work: STC 6899 and CUL reference Bb.^{x} 12.23^{7}F.

It should be noted however that there is no printed date in this Octavo work, and on first principles it would seem strange that there are no other extant titles for a space of two or three years, until 1610 saw a surge of glorious monumental Eton Greek printing bursting upon the scholarly world.

A further puzzle is that the printer, Ionnes Norton, so clearly stated on the title-pages of the major works, was not in fact the real printer of these noble Eton-printed Greek volumes. For the John Norton entry in A Dictionary of Printers.... 1557-1640, under the general editorship of R. B. McKerrow specifically states otherwise, as follows:-

> NORTON (JOHN), senior, bookseller and (?) printer in London, Eton and Scotland, 1586-1612; London: St. Paul's Churchyard. Son of Richard Norton of Billingsley, Salop, and nephew of William Norton, stationer of London (1561-93), to whom he was apprenticed for eight years from January 8th, 157$\frac{7}{8}$ [Arber, ii. 82]. John Norton took up his freedom on July 18th, 1586 [Arber, ii. 698], and rapidly rose to the highest position in the Company, being admitted to the Livery on July 1st, 1598, and being Master in the years 1607, 1611 and 1612. John Norton began business as a bookseller in St. Paul's Churchyard, and was one of the largest capitalists in the trade, besides being the publisher of some of the most important books of the day. His shop was resorted to by the chief book collectors and literary men, and he made regular visits to the Frankfort Fair. He was for a time in partnership with his cousin Bonham Norton [*q.v.*], and John Bill. About 1587 he set up a bookselling business in Edinburgh, and in 1589 obtained, with Andro Hart, the privilege of importing books free of custom, with a further licence to the same effect in June, 1591 [*Reg. P.C. Scot.*, iv. 439; Lee, App. x]. From a passage in Calderwood's *History*, v. 77, it would appear that Norton was living in Edinburgh in 1590; and in February, 1592, he appeared in person before the Town Council to answer a charge preferred against him and his servant Edmond Wats, by seven Edinburgh booksellers, of having usurped the liberty of the burgh by retailing books in "ane oppin chalmer upoun the foregaitt," and they were ordered to desist from selling "in smallis" [Lee, *Add. Mem.* App. lxxi]. After the death of Wats, about 1596, Norton gave up his Edinburgh business, and sold the books and debts to Edward Cathkin and Andro Hart, booksellers there [Calderwood, v. 511]. It has generally been supposed that John Norton was a printer as well as a bookseller, and it is true that in 1603 he was appointed King's printer in Hebrew, Latin and Greek, but on examination the books that bear his imprint are found to have been printed for him by Melchisidec Bradwood and his partners at the Eliot's Court printing office in the Old Bailey. Evidence of this is shown in the splendid edition of Abraham Ortelius' *Theatrum Orbis Terrarum*, 1606, the titlepage of which states distinctly that it was printed "by John Norton, printer to the King's most excellent Majesty in Hebrew, Greeke and Latin," but the colophon of which runs, "London, Printed *for* John Norton and John Bill," and the internal evidence proves it to have been printed at the Eliot's Court Press. So too with Sir Henry Savile's edition of Chrysostom. Each volume bears the imprint "Excudebat Joannes Norton," but we know that Sir Henry Savile obtained the Greek type from Moret, the Antwerp printer, and an examination of the volumes proves that the rest of the type, initials, ornaments and devices were those of Melchisidec Bradwood, who took the necessary workmen down with him to Eton for the purpose of printing the work, which was published by John Norton. Norton died in November, 1612, during his third term of office as Master of the Company. By his will which was proved on January 12th, 161$\frac{2}{3}$, he left a sum of £1,000 to the Company of Stationers to buy lands, the income from which was to be lent to poor young men of the Company. The money was laid out in the purchase of houses in Wood Street, which now produce a considerable rental and form part of the endowment of Stationers' School [Arber, v. lxiii]. John Norton left his lands in Shropshire to provide bequests to his nephew Leonard Norton, son of his brother Richard, and to Lucy and Thomas Wight, the children of Thomas Wight, draper. He made his cousin Bonham Norton his executor [Plomer, *Wills*, pp. 45-7].

Derek Nuttall in his Reading University Thesis (see Bibliography) gives on page 115 additional details about Melchisidec Bradwood's printing career in London and at Eton.

John Norton is also recorded in the Dictionary of National Biography, and it should be added that his name is still remembered annually by the present-day Worshipful Company of Stationers and Newspaper Makers, because at the time of his death in 1612 during his third Mastership he bequeathed (inter alia) a sum of money so that on Ash Wednesday in each February for time to come, cakes, wine and ale would be provided in Stationers Hall, followed by a special service.

My wife and I had the pleasure of attending this Ash Wednesday commemoration in 1988, firstly at Stationers Hall in Ave Maria Street near the top of Ludgate Hill and then at a special service in the crypt of nearby Saint Pauls Cathedral, that is to say in the old chapel of Saint Faith under Saint Pauls, where John Norton himself lies buried. We liverymen walked two by two in procession, through the 'yuppie' City of London crowds, the short distance from Stationers' Hall to the south-west entrance to the crypt for this memorable:

ORDER OF SERVICE

Being assembled, all sit for

THE INTROIT

Ave Verum Corpus ... ... *Anon.*

(Sung by Freeman Alan Anderson)

Then shall the Chaplain welcome the people who will stand :

BELOVED in Christ Jesus, we are met together in this Chapel of St Faith-under-St Paul's, upon Ash Wednesday, to attend Divine Worship, a Bidding Prayer and Sermon, in accordance with the will of John Norton, Alderman of London and Master of the Stationers' Company, who departed this life in the year of our Lord sixteen hundred and twelve.

Let us then commemorate John Norton and all other our benefactors, giving thanks for their foresight and generosity, and pray for this our Company.

So: with Sir Henry Savile, the originator; with John Norton the publisher or trade distributor; and with MELCHISIDEC BRADWOOD and his craftsmen as the practical printers, there resulted eight superb, folio volumes of the works of Saint John Chrysostom. They are all listed under STC 14629 and I have lovingly inspected all eight (in June and in December 1987) in the British Library's North Library (under BL reference c.78.g). Page size is nearly 14" deep (c. 35 cms) by nearly 9" wide and 3" thick. They are leather-bound, the spines with string-courses, plus red-patches gold-lettered 'CHRYOSTOMI OPERA' and then TOM I through to VIII.

These eight impressive volumes record the teachings of John of Antioch who in death was canonised as Saint Chrysostom (meaning 'golden-mouthed') at Constantinople in A.D.438. Trained in oratory for a legal career, he turned to Christian work and teaching, and became Archbishop of Constantinople, an extremely important ecclesiastical position after Constantine, the first Christian Roman Emperor, had moved his capital there. It was this same Constantine the Great who had first been declared Emperor by the Ninth Legion in York, after the death of his father Constantius (Chlorus) in my home City in AD 306. In Constantinople, about a hundred years later, Archbishop John supported the poor and the suffering and declared that holy lives should be led by those in responsible positions. For this plain speaking he was eventually banished into Asia Minor, where he continued his powerful preaching.

To this day in The Book of Common Prayer's Order of Morning Service, his is the final prayer that is said Sunday by Sunday, quoting the New Testament timeless words:
"when two or three are gathered together in Thy Name..."
I reproduce this prayer at the end of this Chapter.

Vol I, or TOMOS PROTOS, of Chrysostum's Opera is the only volume with an engraved title-page: very elaborate, including two birds at top left and right (see facsimile). Also reproduced is the Title-Page-Verso showing the 'Honi Soit' cut at the top and a larger one below, which includes two princesses at lower left and right. The next facsimile is the A signature with a large X initial letter. This goes A/A2/A3 with 3 'plain', making the beginning of the long series of 12 page gatherings. I further show the final printed Volume I folio "1086", and my last Volume I facsimile is the final page with "Errata typographica" followed by "Etonae, Excusum in Collegio Regali, MXCX" i.e. 1610.

All in all Volume I is a substantial folio work of 91 x 12 page gatherings totalling 1,092 pages, though actually the book make-up has a 4 page gathering at the very front and a final 8 pager at the back.

Volume II, or TOMOS DEUTEROS, is of similar page size and book make-up, comprising a total of 940 pages. This includes a design of two twining snakes on the title-page, and bears the date of 1610 at front and at back.

Volume III, or TOMOS TRITOS, comprises 898 pages, with the year 1611 on the title page, but perhaps forgetfully perpetuating "MDCX" on the final printed folio 894.

Volume IV, or TOMOS TETARTOS, comprises 924 pages with 1612 consistently shown as the date at front and back.

Volume V, or TOMOS PEMPTOS, was still larger being another Compositor's eye-strainer and Pressman's back-breaker - of 1,004 pages. Strangely Volume V reverts to the year 1611 at both front and back, whereas its presumed predecessor Volume IV is consistently 1612.

In these early printing studies I am blessed with the help of skilled advisers who through this series have alerted me at draft stage to the usual highly-amusing pattern of authors' howlers. Upon my querying this Volume IV/Volume V dating inconsistency, my manuscript North Library notes though voluminous were such that 'TOMOS PEMPTOS' (Fifth) at first looked like MEMPTOS which could be construed "to be blamed", hence suspect. Thus I fool-hardily attempt my first and last Greek joke that "the dates printed in TOMOS PEMPTOS may be MEMPTOS". (Although I had a 'good' grading in matric latin to qualify me for Cambridge entry, I admit openly that Greek is "Greek to me").

With the defence that the great Shakespeare deliberately introduced moments of lighter relief into his tragedies, I revert, and not before time, to my British Library examination of Volume VI or, TOMOS EKTOS, another 'back-breaker' for the working printers with a total of 1,012 pages in each and every copy. Strangely the dating at front and at back here reverts also to 1611.

Volume VII, or TOMOS (H)EBDOMOS, does however show consistent 1612 dating, with only a slightly fewer number of pages, namely 952.

The final Volume VIII, or TOMOS OGDOOS, ("as in Co-op": Bernard Barr), lacks a final colophon but is clearly 1612 in latin on the title page. The text composition, and so the folioing of this final volume is much more complex than the previous seven books, but my count of main text, double-columned notes, index etc. gives me a total of 999 printed pages.

STC[2] summarises the variants of 14629 as follows:-

14629 **John**, *Chrysostom, Saint. Του ἐν ἁγίοις πατρὸς ἡμῶν Ιωαννου τοῦ Χρυσοστόμου τα ευρισκομενα.* [*Ed.* Sir H. Savile.] 8 vols. fol. *Etonæ*, [*M. Bradwood f.*] *J. Norton*, 1610 (1611, 1612.) L.

Without prelims. except for the tpp. Tp dates vary: vols. 1, 2: 1610; vols. 3, 5, 6: 1611; vols. 4, 7, 8: 1612, all in arabic numerals except for vol. 8, which has: 'cIↄ. Iↄc. XII.' Colophons dated: vols. 1–3: 1610; vols. 5, 6: 1611; vols. 4, 7: 1612; vol. 8 dated 1613 on ¶¶¶4^v. At the end of vol. 8 L has bd. in quires Iiii–Kkkk of vol. 4 and the over-all contents list from vol. 1 of 14629a.

14629.5 — [A variant.] L^2(tpp dated: vols. 2, 3, 5, 6: 1611; vols. 1, 4, 7, 8: 1612).O^{10}(tpp dated: vols. 1–3, 5, 6: 1611; vols. 4, 7, 8: 1612).

(Formerly 14629b) L^2 has at least the vol. 1 prelims. of 14629a. O^{10} has a complete set of the prelims. for 14629a bd. in vol. 8.

14629a — [Anr. issue, w. new prelims. for each vol., w. tpp dated 'cIↄ. Iↄc. XII [or XII]' and an added engr. tp in vol. 1, w. title:] S. Joannis Chrysostomi opera Græcé. *Etonæ, J. Norton*, 1613. L.O.C.D.E^2.+; F.HN. CU.HD.N.+

(Formerly also 14629c) Vol. 1 prelims. also include dedic., epistle, and over-all contents list; prelims. for the other vols. include only individual contents list.

The copy in St. Mark's Cathedral, Venice, has a pr. dedic. from Savile to Marcantonio Memmo, Doge, and to the Senate of Venice.

14629b — Now = 14629.5.

14629c — [Anr. issue, dated:] 1612. = 14629a.

— A cōpendious treatise. 1542. *See* 14640.

I show at the end of this chapter, firstly courtesy of The British Library London, a range of reduced-size facsimiles from the STC 14629 first editions of these eight impressive volumes, including all the title-pages and all the final colophons.

Secondly, courtesy of York Minster Library, I include facsimiles of all the variant STC 14629a title pages. I also show, at the beginning of this latter sequence, the engraved title-page covering the whole eight volumes, which appears right at the beginning of Volume I, and which is even more magnificent in its full folio size of circa 12½" x nearly 8".

It will be seen that this series title-page for the second edition of eight volumes is dated: 1613, this being one year after John Norton's death.

Paul R Quarrie MA, FSA, Eton College Librarian, kindly informs me that "The printer's copy for the Eton Chrysostom is in Bodley (MS Auct.E.3 1-16, 4.1-6 - codices miscellanei 51 in H.O.Coxe's catalogue of Greek Manuscripts in the Bodleian)".

There are other extant Eton-printed works in Greek type in this same short span of years, which I have also inspected at the British Library in London, namely:-

(a) STC 14622 and BL Reference 3670 bb 26(2)

IONNIS METROPOLITANI EVCHAITENSIS

also avowedly printed by John Norton - in Eton in 1610. This is a smaller work of about 9-1/4" x 7" page size, and of 90 pages in total, with 8-pager gatherings, and with the same two twining snakes on the title-page, as with DIONYSIUS already described.

(b) STC 26065 and BL Reference 586.e 25(1)

XENOPHONTIS again with the twining snakes

and Eton-printed three years later in 1613. A work of 390 pages with smaller page size about 7½" x 5½". The final colophon page is partly destroyed on this British Library copy, but 1613 and Eton printing are consistent and clear, though with no 'John Norton' claim at either front or back.

(c) SANCTI GREGORII NAZIANZENI: I found this a little puzzling in the sense that Clough lists two Eton-printed editions, namely:-

STC 12346 of 1610 Quarto with John Norton imprint and
STC 12347 of 1615 Octavo without John Norton being names.

The former I located under British Library Reference 1124 f.31, in correct page size of about 9-1/4" x 7", and with John Norton duly named (see Facsimile). This is again a substantial work of 262 pages.

Additionally I called forth British Library Reference 3670 bb. 26(1) believing that this would prove to be the smaller-page-size 1615 edition. In the event it proved to be a seemingly exact duplicate of the quarto 1610 edition, so I have yet to examine this 1615 work which to my knowledge is the latest extant work from this very special, short-lived Eton College Press.

My eagerness to locate this 1615 Eton work is all the greater because W H Allnutt in the Eton section (pages 276-278) of his series of articles on the "English Provincial Press" in the <u>Bibliographica</u> Journal (open shelves British Library's North Library) records this 1615 Eton title as

> 5. *Γρεγορίου Ναζανζηνοῦ τοῦ θεολόγου λόγος εἰς τὰ θεοφάνια εἴτ' οὖν γενέθλια τοῦ σωτῆρος.* [Printer's device with initials 'M. B.'] *Etonae Excusum in Collegio Regali Anno* cIↄ Iↄc xv. 8°.

The words in parenthesis are of riveting interest, because I submit that this "Printers' device with initials M.B." will add proof to the Eton printing having been in fact the work of Melchisidec Bradwood.

Whereas John Norton, the stated Eton printer and actual-publisher, died in 1612 (in November, during his third term of office as Master of the London Stationers Company), the entry in the same <u>Dictionery of Printers</u> for the actual Eton printer, Melchisidec Bradwood states that he, Bradwood, "remained in Eton till his death and was the printer of other Greek works. He died between the 6th and the 30th June 1618, and by his will left to his wife all his stock in the Company of Stationers".

The John Norton death date of 1612 is interesting as marking also the completion date of the substantial series of eight Chrysostom volumes. The 1618 date of Bradwoood's death in Eton is however three years after the latest extant Eton title as listed by Clough.

The high standard of workmanship of this short-lived Eton Press is greatly to be admired. A recent assessment as to the technical difficulty of setting and proof-reading Greek type is contained in an article by the typographical historian Hugh Williamson, in the May 1987 issue of the Printing Historical Society Bulletin which he edits, namely:-

> These [slightly earlier] Greek types were a marvellous imitation of cursive manuscript, but they can never have been easy to read and they must have been abominably laborious to set. The number of variant characters and elaborate ligatures were immense, and in addition to these there were the customary accents and breathings, of which in a Greek fount of the twentieth century there were over 120 combinations with the seven lower-case vowels alone. Setting such founts accurately and not too slowly must have been a special skill then as it would be now........

The sheer output of this Eton press is also noteworthy and warrants even an incomplete summary, thus:-

			pages
Chrysostomi	I		1,092
	II		940
	III		898
	IV		924
	V		1,004
	VI		1,012
	VII		952
	VIII	circa	999
IONNIS			90
XENOPHONTIS			390
GREGORY 1610			262
Total Eton Pages			8,563

If even an average of 100 copies of each work were produced, this would of course have involved an output in excess of 800,000 large folio printed pages - a back-straining job for the hand-press inkers and pullers (judging by my occasional two-hourly hand-press stints at York's Castle Museum).

The sustained inspiration and scholarly oversight (and perhaps difficult proof-reading) by Sir Henry Saville is also to be commended. I am glad that his well-deserved monument in the chapel of Merton College Oxford, includes not only Tacitus and Euclid, but also the lesser-known but important Saint John Chrysostom of Antioch and Constantinople.

My wife and I had the great pleasure of visiting Eton College on Easter Saturday 1988 (courtesy of Mr Broughton and Mr Christy Miller), including the Museum of Eton Life, housed in the vaulted King's Cellar, under the College Hall, which museum the Queen Mother graciously opened in April 1986. Here is displayed a copy of the 1610 Eton-printed Ionnis Metropolitani Evchaitensis, already described; and additionally the 1613 series title-page of the second, or composite, edition of Chrysostom, as donated to the Eton College Library by Sir Henry Savile himself (ECL Fe 21).

Mr Paul Quarrie, Eton College Librarian, kindly informs me that the actual location of the early Eton press was in what is now called Savile House, which lies close to the main school courtyard on the non-Chapel side, fronting the present Slough Road, as shown (courtesy of Eton College) on the plan below:

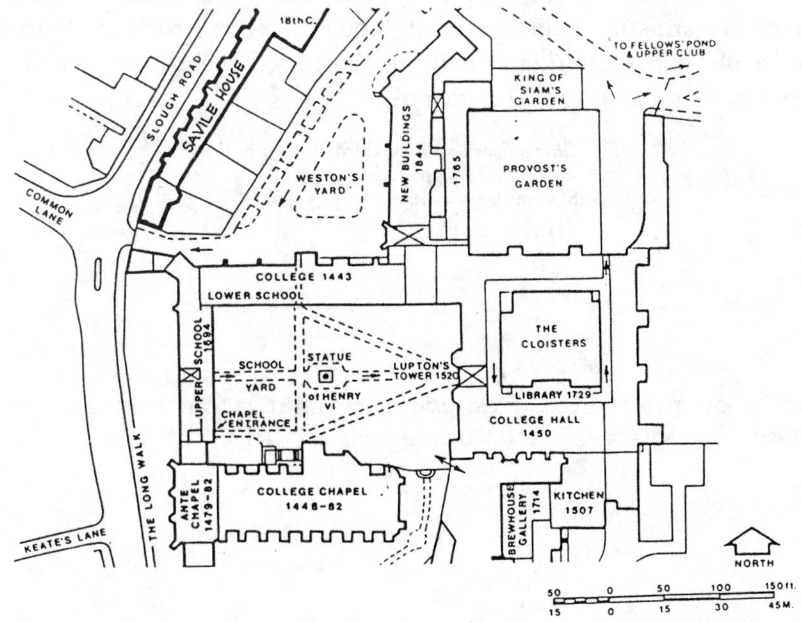

I am also informed that two of the individual volumes of Eton-printed Greek texts were designated for School teaching use. I understand that Sir Henry Savile expended the considerable sum of £6,000 on his Eton printing enterprise. Subsequent to 1615 compilation of text books was continued by former Eton and King's College Cambridge students, but these were printed in London, and Eton printing did not resume until 1730, by Joseph Pote in the High Street, who both printed and published, being succeeded by his son Thomas Pote. Their highly successful Eton Latin Grammar was used by many schools and in its heyday was reprinted every year.

The Eton Greek type used in the monumental eight volumes of Chrysostom, together with the matrices (from which it had been hand-cast with whirling arms) was in 1619 given by Sir Henry Savile to Oxford University to help the establishment of the University's own printing press. Harry Carter writes on p 25 of his History of the Oxford University Press, Volume I (1975):-

> The University was brought a step nearer to a printing business of its own by an acquisition of printing type in 1619. The Vice-Chancellor entered in his accounts for the year ending 21 July 1620 two items for 'carriage to Oxford of the Greek types given to the University by Sir Henry Savile'. The compiler of a list of Savile's benefactions writes in 1622, as though the University already owned a press: 'Officinam Universitatis Typographicam literis graecis innumeris, literarumque matricibus (monimento insigni) ditavit.'[1] The fount, it is easy to see, is one with which Savile printed the works of St. Chrysostom at Eton in 1610–13 in six folio volumes. It is a big type, suitable for texts, whereas Barnes and his successors had only a small Greek, fit for footnotes. The matrices, in number 319, of which 153 are for ligatured letters and contractions, are still at the Press with added ligatured sorts made to the order of Fell.[2] Cambridge University borrowed them in 1629 and had type cast for printing a Testament.[3]
>
> 1 *Ultima linea Savilii* (Oxford, 1622), p. [iii].
> 2 Morison, *Fell*, pp. 102–3.
> * 3 Madan, *Oxford Books*, ii, pp. 517–20.

* Madan adds on p xiii of his preface that this was for "the first Cambridge Greek New Testament of 1632".

Nicolas Barker, has on pages 10-11 of his The Oxford University Press and Spread of Learning (1978) a short section of which items 22 to 25 relate to Sir Henry Savile and his Greek type, as follows:-

THE ACQUISITION OF PRINTING TYPES

21. Charles Butler. *Oratoriae libri duo.* Oxoniae, excudebat Guilielmus Turner, impensis authoris, 1629.

There had been little advance in the typographic equipment of the Oxford printers since Barnes's time. An interesting specimen of it comes in Butler's tract on oratory. Making the classic point that all sentences depend on variety, he analyses the *distinctiones*, which can be a difference of punctuation or letters; the latter differ in kind (Roman or Italic) or shape (capitals or lower-case letters), and in size. He illustrates the point with a range of type sizes, and a specimen of Roman, Italic, and 'English' or Black-letter. It shows the normal types available in Oxford at the time.

22. Portrait of Sir Henry Savile (1549–1622), attributed to Marcus Gheeraerts (Bodleian Library).

23. Matrices of the Greek type given to the University by Savile.

24. *Ultima linea Savilii sive in obitum clarissimi Domini Henrici Savilii.* Oxoniae, excudebant Johannes Lichfield et Jacobus Short, 1622.

25. *Camdeni insignia.* Oxoniae, excudebant Johannes Lichfield et Jacobus Short, Academiae typographi, 1624.

Sir Henry Savile, Warden of Merton College and later Provost of Eton, was a great benefactor of learning. His most notable achievement was the edition of the works of St. John Chrysostom in eight volumes, printed at the press that he established at Eton in 1610–13. In 1619 he gave his Greek type to the University, and the gift is recorded in *Ultima linea Savilii*, a record of his benefactions with verses to his memory.

The sources of his type, a large Great Primer Greek fitter for texts than Barnes's smaller type, are not clear. Savile had wanted a set of the celebrated *grecs du roi*, cut for François I by Claude Garamont. The type he got was a copy, probably bought at the Frankfurt book fair in 1608 and perhaps originally cut by Pierre Haultin *c.* 1580. The equipment given by Savile was first used in the University's tribute to William Camden (1551–1623), another great benefactor (founder of the chair of history) and scholar. It differs slightly from the type used by Savile at Eton (some sets of punches, matrices, and type do not always match), and has been augmented and adapted since. The bulk of it remains, a relic of the oldest gift of type to the University.

A Prayer of Saint Chrysostom.

ALMIGHTY God, who hast given us grace at this time with one accord to make our common supplications unto thee; and dost promise that when two or three are gathered * together in thy Name thou wilt grant their requests: Fulfil now, O Lord, the desires and petitions of thy servants, as may be most expedient for them; granting us in this world knowledge of thy truth, and in the world to come life everlasting. ***Amen.***

* cf. Matthew 18 'For where 2 or 3 have met together in my name, I am there among them.'

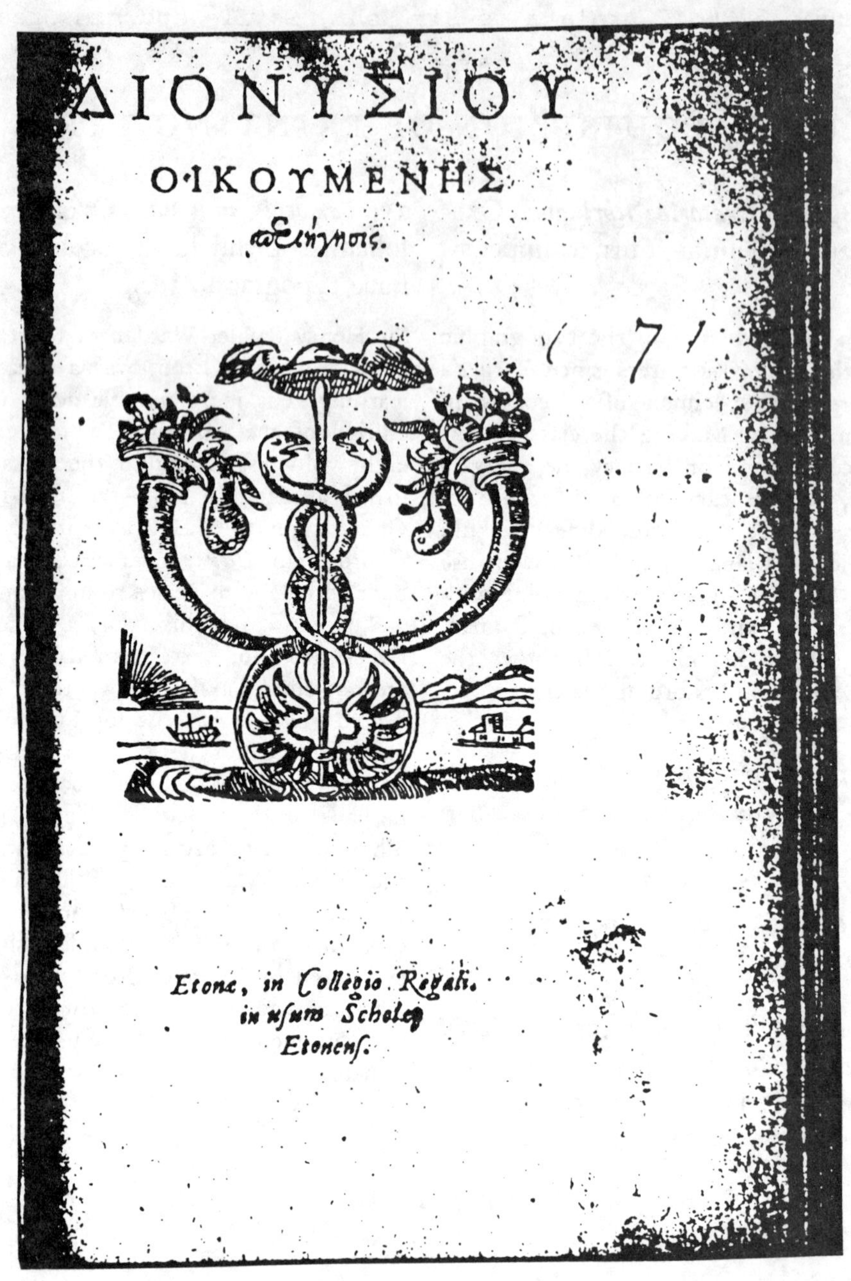
ΔΙΟΝΥΣΙΟΥ

ΟΙΚΟΥΜΕΝΗΣ

περιήγησις.

Etonæ, in Collegio Regali.
in usum Scholæ
Etonens.

STC 6899 DIONYSIUS PERIEGETES

printed at Eton [c.1607?]

Courtesy Cambridge University Library

All 8 Volumes of Chrysostom are STC 14629.

The magnificent originals are c 11-3/4" x 9-1/4" leather bound with old gold-tooling.

ΤΟΥ ἐν ἁγίοις πατρὸς ἡμῶν

ΙΩΑΝΝΟΥ

ΑΡΧΙΕΠΙΣΚΟΠΟΥ

Κωνσταντινουπόλεως τοῦ Χρυσοστόμου

ΤΑ ΕΥΡΙΣΚΟΜΕΝΑ,

Δι' ἐπιμελείας καὶ ἀναλωμάτων

ΕΡΡΙΚΟΥ τοῦ ΣΑΒΙΛΙΟΥ

ἐκ παλαιῶν ἀντιγράφων ἐκδοθέντα.

ΤΟΜΟΣ ΠΡΩΤΟΣ.

ETONAE,

In Collegio Regali,

Excudebat IOANNES NORTON in Graecis, &c. Typographus Regius.

ANNO CIↃ IↃC X.

Cum privilegio.

CHRYSOSTOMI OPERA

Volume I or Tomos Protos

Courtesy British Library London

ΙΑΚΩΒΩ
ΤΩ
ΕΝΔΟΞΟΤΑΤΩ
ΚΑΙ
ΜΕΓΙΣΤΩ
ΜΕΓΑΛΗΣ ΒΡΕΤ-
ΤΑΝΙΑΣ,
ΦΡΑΓΚΙΑΣ
ΚΑΙ
ΙΟΥΕΡΝΙΑΣ
ΒΑΣΙΛΕΙ
ΔΩΡΟΝ.

Volume I continued

Courtesy British Library London

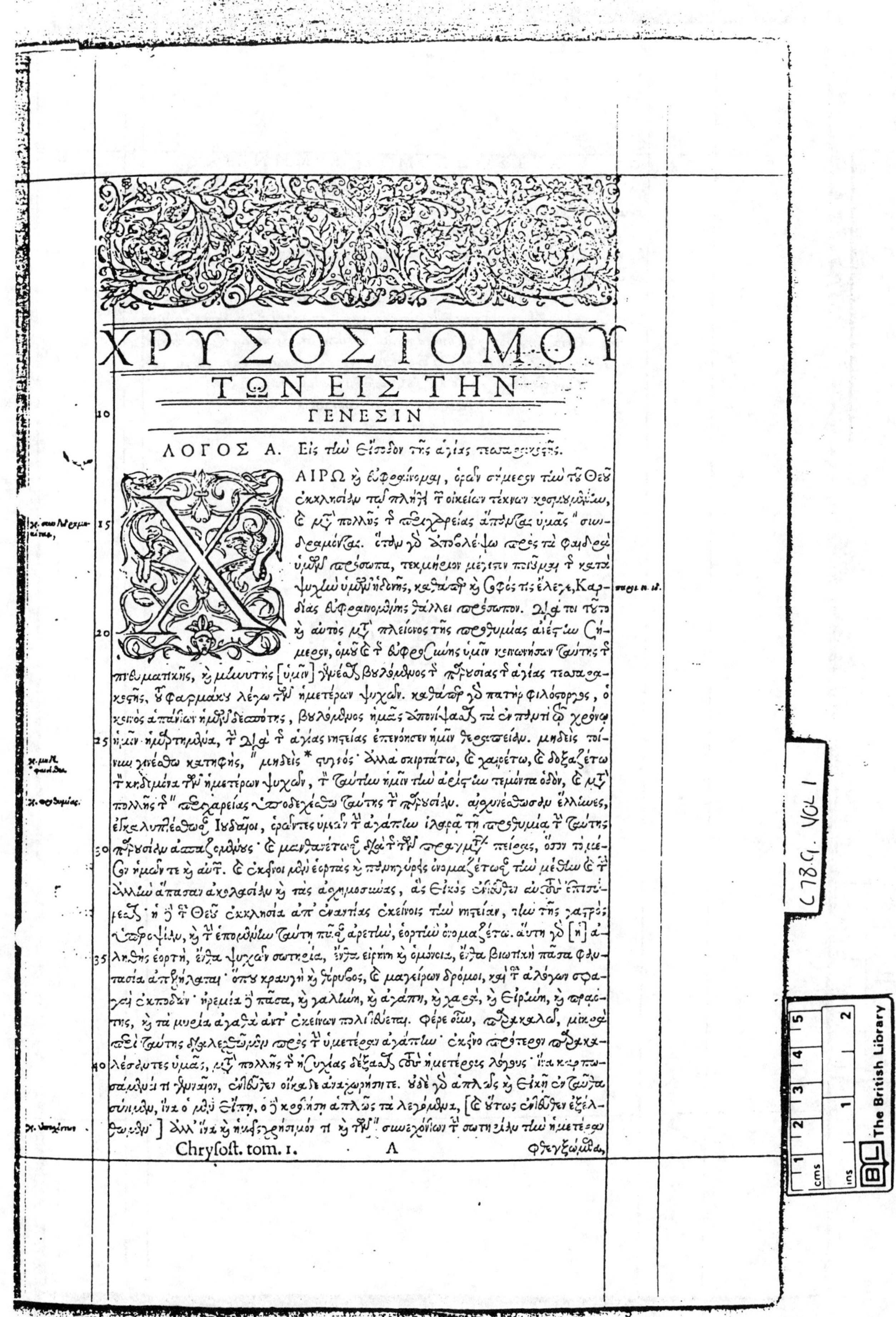

ΧΡΥΣΟΣΤΟΜΟΥ
ΤΩΝ ΕΙΣ ΤΗΝ
ΓΕΝΕΣΙΝ

ΛΟΓΟΣ Α. Εἰς τὴν εἴσοδον τῆς ἁγίας τεσσαρακοστῆς.

ΧΑΙΡΩ καὶ εὐφραίνομαι, ὁρῶν σήμερον τὴν τοῦ Θεοῦ ἐκκλησίαν τῷ πλήθει τῶν οἰκείων τέκνων κοσμουμένην, καὶ μετὰ πολλῆς τῆς περιχαρείας ἅπαντας ὑμᾶς συνδραμόντας. ὅταν γὰρ ἀποβλέψω πρὸς τὰ φαιδρὰ ὑμῶν πρόσωπα, τεκμήριον μέγιστον ποιοῦμαι τῆς κατὰ ψυχὴν ὑμῶν ἡδονῆς, καθάπερ καὶ σοφός τις ἔλεγε, Καρδίας εὐφραινομένης θάλλει πρόσωπον. διά τοι τοῦτο καὶ αὐτὸς μετὰ πλείονος τῆς προθυμίας ἀνέστην σήμερον, ὥστε καὶ τῆς εὐφροσύνης ὑμῖν κοινωνῆσαι ταύτης τῆς πνευματικῆς, καὶ μηνυτὴς [ὑμῖν] γενέσθαι βουλόμενος τῆς παρουσίας τῆς ἁγίας τεσσαρακοστῆς, τοῦ φαρμάκου λέγω τῶν ἡμετέρων ψυχῶν. καθάπερ γὰρ πατὴρ φιλόστοργος, ὁ κοινὸς ἁπάντων ἡμῶν δεσπότης, βουλόμενος ἡμᾶς ἀπονίψασθαι τὰ ἐν παντὶ τῷ χρόνῳ ἡμῖν ἡμαρτημένα, τὴν διὰ τῆς ἁγίας νηστείας ἐπενόησεν ἡμῖν θεραπείαν. μηδεὶς τοίνυν γινέσθω κατηφής, μηδεὶς σκυθρωπός· ἀλλὰ σκιρτάτω, καὶ χαιρέτω, καὶ δοξαζέτω τὸν κηδεμόνα τῶν ἡμετέρων ψυχῶν, τὸν ταύτην ἡμῖν τὴν ἀρίστην τεμόντα ὁδόν, καὶ μετὰ πολλῆς τῆς περιχαρείας ὑποδεχέσθω ταύτης τὴν παρουσίαν. αἰσχυνέσθωσαν Ἕλληνες, ἐγκαλυπτέσθωσαν Ἰουδαῖοι, ὁρῶντες ὑμῶν τὴν ἀγάπην ἱλαρᾷ τῇ προθυμίᾳ τὴν ταύτης παρουσίαν ἀσπαζομένους· καὶ μανθανέτωσαν διὰ τῶν πραγμάτων πείρας, ὅσον τὸ μέσον ἡμῶν τε καὶ αὐτῶν. καὶ ἐκεῖνοι μὲν ἑορτὰς καὶ πανηγύρεις ὀνομαζέτωσαν τὴν μέθην καὶ τὴν ἄλλην ἅπασαν ἀκολασίαν καὶ τὰς ἀσχημοσύνας, ἃς εἰκὸς ἐντεῦθεν αὐτοὺς ἐπιτελεῖσθαι· ἡ δὲ τοῦ Θεοῦ ἐκκλησία ἀπ' ἐναντίας ἐκείνοις τὴν νηστείαν, τὴν τῆς γαστρὸς ὑπεροψίαν, καὶ τὴν ἑπομένην ταύτῃ πᾶσαν ἀρετήν, ἑορτὴν ὀνομαζέτω. αὕτη γὰρ [ἡ] ἀληθὴς ἑορτή, ἔνθα ψυχῶν σωτηρία, ἔνθα εἰρήνη καὶ ὁμόνοια, ἔνθα βιωτικὴ πᾶσα φαντασία ἀπελήλαται· ὅπου κραυγὴ καὶ θόρυβος, καὶ μαγείρων δρόμοι, καὶ τῶν ἀλόγων σφαγαὶ ἐκποδών· ἠρεμία δὲ πᾶσα, καὶ γαλήνη, καὶ ἀγάπη, καὶ χαρά, καὶ εἰρήνη, καὶ πρᾳότης, καὶ τὰ μυρία ἀγαθὰ ἀντ' ἐκείνων πολιτεύεται. φέρε οὖν, παρακαλῶ, μικρὰ περὶ ταύτης διαλεχθῶμεν πρὸς τὴν ὑμετέραν ἀγάπην· ἐκεῖνο πρότερον παρακαλέσαντες ὑμᾶς, μετὰ πολλῆς τῆς ἡσυχίας δέξασθαι τοὺς ἡμετέρους λόγους· ἵνα καρπωσάμενοί τι χρήσιμον, ἐντεῦθεν οἴκαδε ἀναχωρήσητε. οὐδὲ γὰρ ἁπλῶς καὶ εἰκῇ ἐνταῦθα σύνιμεν, ἵνα ὁ μὲν εἴπῃ, ὁ δὲ κροτήσῃ ἁπλῶς τὰ λεγόμενα, [καὶ οὕτως ἐντεῦθεν ἐξέλθωμεν] ἀλλ' ἵνα καὶ ἡμεῖς χρήσιμόν τι καὶ τῶν συνεχόντων τὴν σωτηρίαν τὴν ἡμετέραν

Volume I continued

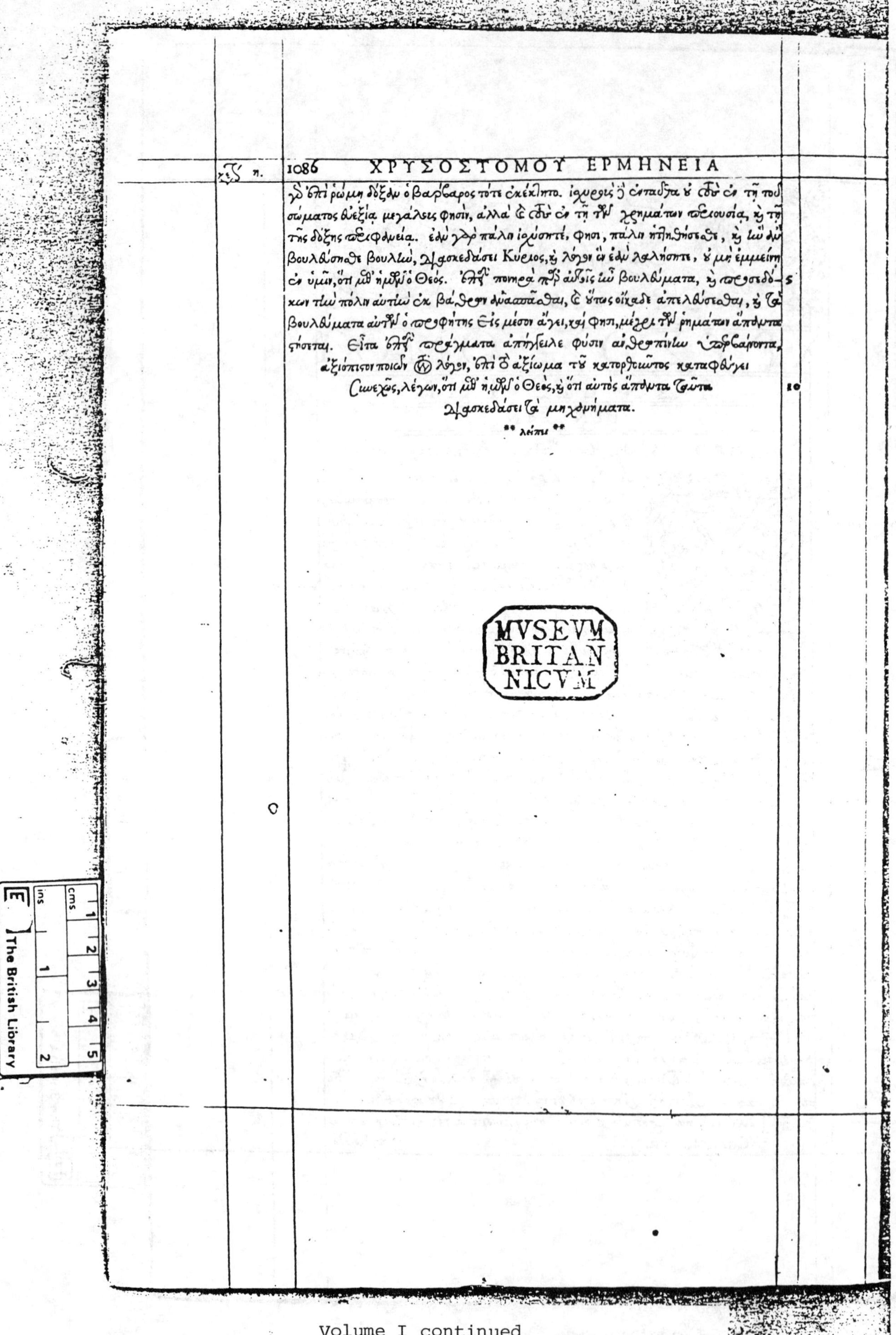

γὰρ ἐπὶ ῥώμῃ δόξαν ὁ βάρβαρος τότε ἐκέκτητο. ἰσχυροὺς δὲ ἐνταῦθα οὐ τοὺς ἐν τῇ τοῦ σώματος δεξιᾷ μεγάλους φησίν, ἀλλὰ τοὺς ἐν τῇ τῶν χρημάτων περιουσίᾳ, καὶ τῇ τῆς δόξης περιφανείᾳ. ἐὰν γὰρ πάλιν ἰσχύσητε, φησί, πάλιν ἡττηθήσεσθε, καὶ ἣν ἂν βουλεύσησθε βουλήν, διασκεδάσει Κύριος, καὶ λόγον ὃν ἐὰν λαλήσητε, οὐ μὴ ἐμμείνῃ ἐν ὑμῖν, ὅτι μεθ' ἡμῶν ὁ Θεός. ἐπειδὴ πονηρὰ περὶ αὐτοῖς ἦν βουλεύματα, καὶ προσεδόκων τὴν πόλιν αὐτὴν ἐκ βάθρων ἀνασπάσασθαι, καὶ οὕτως οἴκαδε ἀπελεύσεσθαι, καὶ τὰ βουλεύματα αὐτῶν ὁ προφήτης εἰς μέσον ἄγει, καί φησι, μέχρι τῶν ῥημάτων ἅπαντα στήσεται. Εἶτα ἐπειδὴ πράγματα ἀπήγγειλε φύσιν ἀνθρωπίνην ὑπερβαίνοντα, ἀξιόπιστον ποιῶν τὸν λόγον, ἐπὶ τὸ ἀξίωμα τοῦ κατορθοῦντος καταφεύγει συνεχῶς, λέγων, ὅτι μεθ' ἡμῶν ὁ Θεός, καὶ ὅτι αὐτὸς ἅπαντα ταῦτα διασκεδάσει τὰ μηχανήματα.

** λείπει **

Volume I continued

Courtesy British Library London

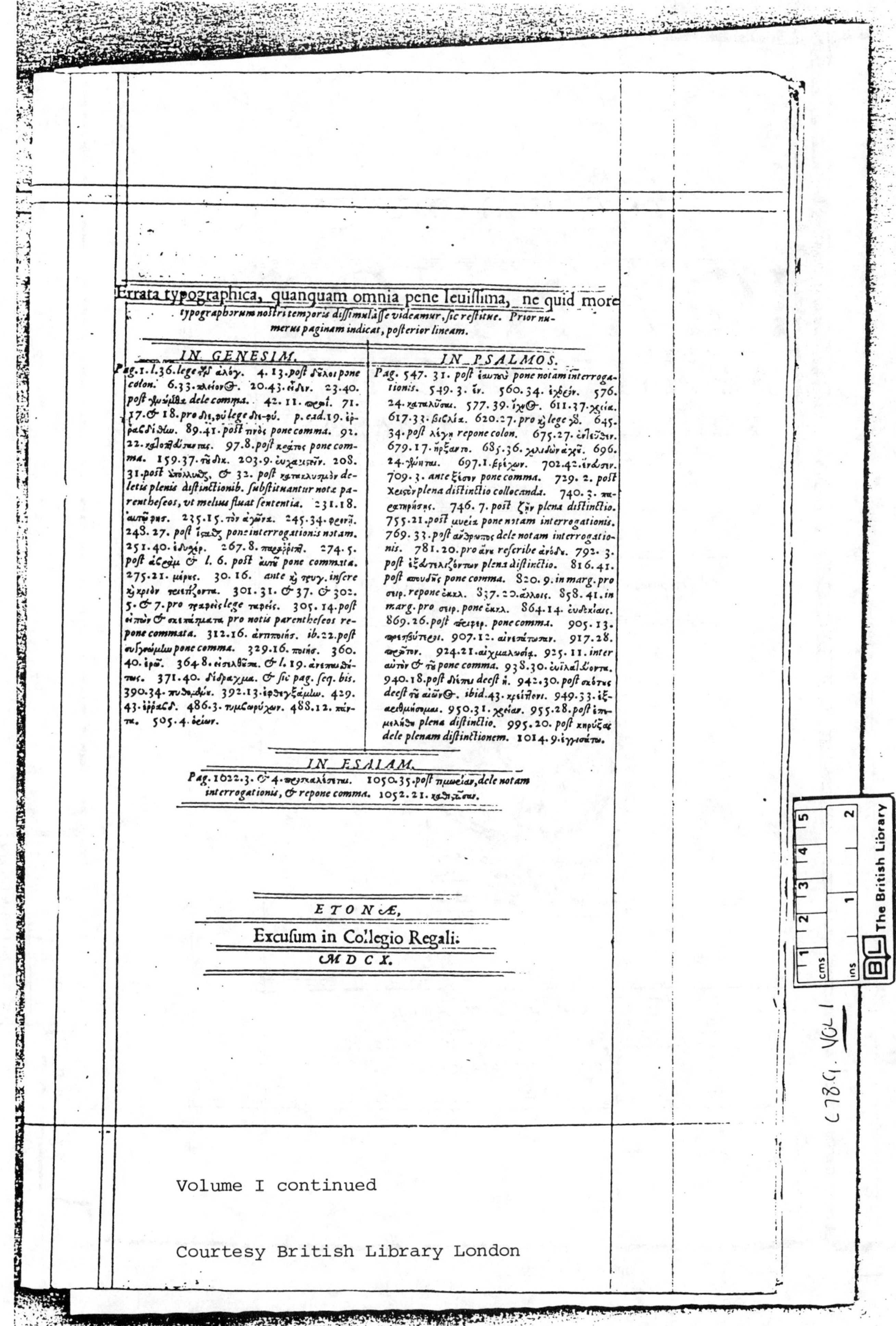

Errata typographica, quanquam omnia pene leuissima, ne quid more typographorum nostri temporis dissimulasse videamur, sic restitue. Prior numerus paginam indicat, posterior lineam.

IN GENESIM.

Pag. 1. l. 36. lege τῆς ἀλόγ. 4. 13. post δοῦλοι pone colon. 6. 33. πλείονος. 20. 43. εἶδεν. 23. 40. post γυναῖκα dele comma. 42. 11. περί. 71. 17. & 18. pro δι, ού lege δι-ού. p. ead. 19. ἐρραβδισθω. 89. 41. post τινὸς pone comma. 92. 22. κατασταθήσονται. 97. 8. post κράτος pone comma. 159. 37. τοῦ δίκ. 203. 9. εὐχαριστεῖν. 208. 31. post ἀπόλλυσθαι, & 32. post κατακλυσμὸν deletis plenis distinctionib. substituantur notæ parentheseos, ut melius fluat sententia. 231. 18. αὐτῷ φησ. 235. 15. τὸν ἀγῶνα. 245. 34. φθορᾷ. 248. 27. post ἴσασι pone interrogationis notam. 251. 40. ἐδυσχέρ. 267. 8. παρεῤῥίφθη. 274. 5. post ἀβραὰμ & l. 6. post αὐτοῦ pone commata. 275. 21. μέρους. 30. 16. ante καὶ τριγ. insere καὶ χριὸν τεσσαράκοντα. 301. 31. & 37. & 302. 5. & 7. pro τεταρεῖς lege ταρεῖς. 305. 14. post εἰπεῖν & σκεπάσματα pro notis parentheseos repone commata. 312. 16. ἀντιποιήσ. ib. 22. post συγγνώμην pone comma. 329. 16. ποιῆσ. 360. 40. ἐρῶ. 364. 8. εἰσελθοῦσα. & l. 19. ἀνεπαισθήτως. 371. 40. δίδραχμα. & sic pag. seq. bis. 390. 34. πυθομένου. 392. 13. ἐφθεγξάμην. 429. 43. ἐρράβδ. 486. 3. τυμβωρύχων. 488. 12. πάντα. 505. 4. θείων.

IN PSALMOS.

Pag. 547. 31. post ἑαυτοῦ pone notam interrogationis. 549. 3. ἐν. 560. 34. ἐχθρῶν. 576. 24. καταλῦσαι. 577. 39. ἴχνος. 611. 37. χεῖρα. 617. 33. βιβλία. 620. 27. pro καὶ lege γὰρ. 645. 34. post λέγῃ repone colon. 675. 27. ἐνθεῦθεν. 679. 17. ἤρξαντο. 685. 36. χελιδὼν ἀχεῖ. 696. 24. φύονται. 697. 1. βρέχων. 702. 42. ἐνδόσιν. 709. 3. ante ξίφον pone comma. 729. 2. post Χριστὸν plena distinctio collocanda. 740. 3. παρατηρήσης. 746. 7. post ζῆν plena distinctio. 755. 21. post μνεία pone notam interrogationis. 769. 33. post ἄνθρωπος dele notam interrogationis. 781. 20. pro ἄνευ rescribe ἀνόδ. 792. 3. post ἐξακοντιζόντων plena distinctio. 816. 41. post σπουδῆς pone comma. 820. 9. in marg. pro στιχ. repone ἐκκλ. 837. 20. ἄλλοις. 858. 41. in marg. pro στιχ. pone ἐκκλ. 864. 14. εὐδοκίαις. 869. 26. post φείτηρ. pone comma. 905. 13. πρεσβύτεροι. 907. 12. αἰνεσάτωσαν. 917. 28. πρῶτον. 924. 21. αἰχμαλωσίᾳ. 925. 11. inter αὐτὸν & τοῦ pone comma. 938. 30. εὐιλάζοντα. 940. 18. post δίκαιοι deest ἡ. 942. 30. post οὕτως deest τοῦ αἰῶνος. ibid. 43. κρείττονι. 949. 33. ἐξαριθμήσομαι. 950. 31. χεῖραν. 955. 28. post ἐπιμελήθη plena distinctio. 995. 20. post κηρύξαι dele plenam distinctionem. 1014. 9. ἐγγισάτω.

IN ESAIAM.

Pag. 1022. 3. & 4. περιπαλήσεται. 1050. 35. post τιμωρίαν, dele notam interrogationis, & repone comma. 1052. 21. καθιζῆσαι.

ETONÆ,

Excusum in Collegio Regali.

MDCX.

Volume I continued

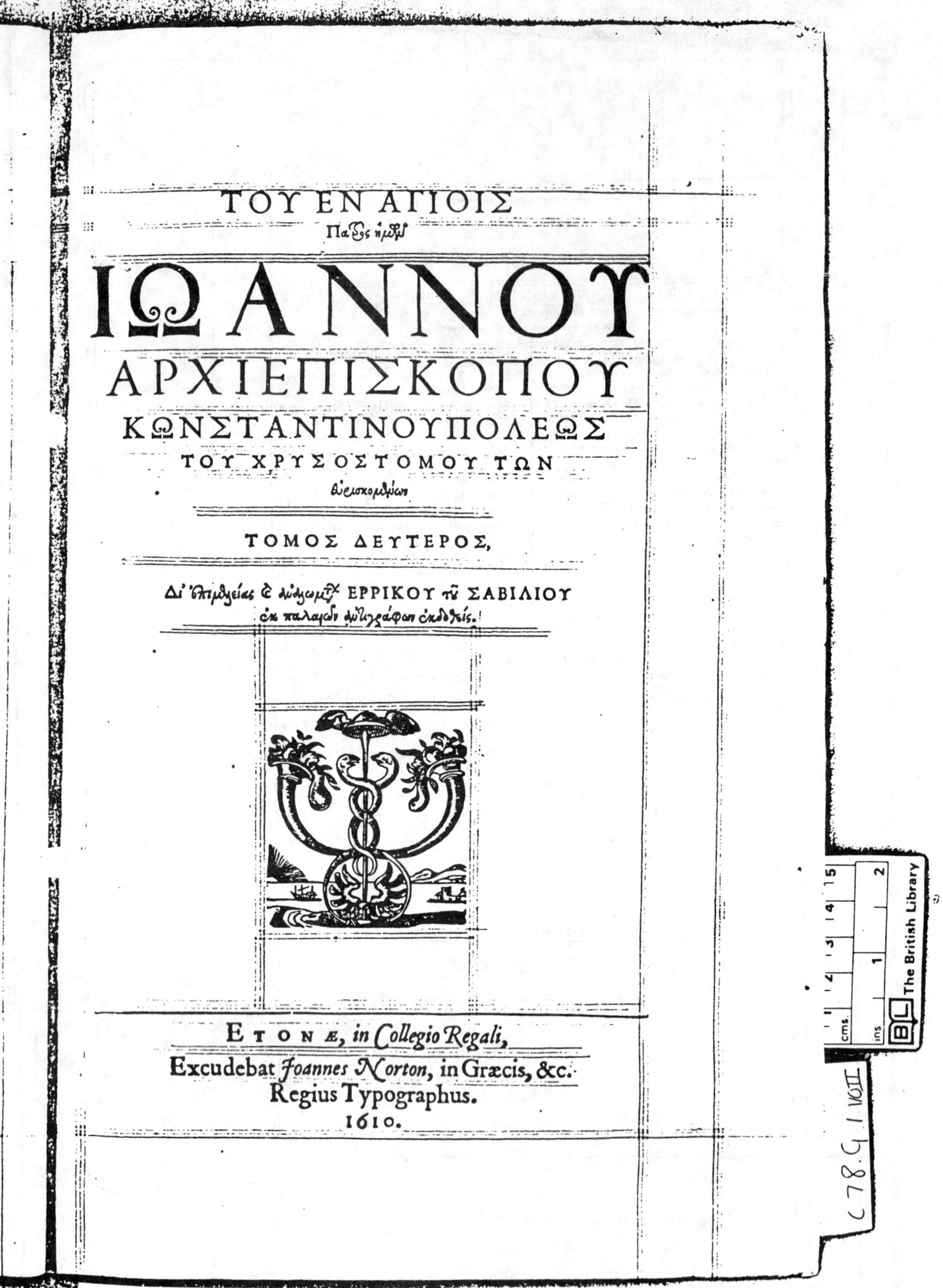

ΤΟΥ ΕΝ ΑΓΙΟΙΣ

Πατρὸς ἡμῶν

ΙΩΑΝΝΟΥ

ΑΡΧΙΕΠΙΣΚΟΠΟΥ

ΚΩΝΣΤΑΝΤΙΝΟΥΠΟΛΕΩΣ

ΤΟΥ ΧΡΥΣΟΣΤΟΜΟΥ ΤΩΝ

εὑρισκομένων

ΤΟΜΟΣ ΔΕΥΤΕΡΟΣ,

Δι' ἐπιμελείας καὶ ἀναλωμάτων ΕΡΡΙΚΟΥ τοῦ ΣΑΒΙΛΙΟΥ
ἐκ παλαιῶν ἀντιγράφων ἐκδοθείς.

ETONÆ, *in Collegio Regali,*
Excudebat *Joannes Norton,* in Græcis, &c.
Regius Typographus.
1610.

Volume II or Tomos Deuteros

Courtesy British Library London

δυνησόμεθα ἀναπνεῦσαι; καὶ γὰρ εἴ τις τὸν μέλλοντα καταπίπτειν ἄνωθεν ἕλκοι (ὅπερ ἐστὶν ἡ ἐλεημοσύνη) ἕτερος δὲ κάτωθεν βιάζοιτο, οὐδὲν ἔσται τι πλέον ἀπὸ τῆς τοιαύτης πάλης, ἢ τὸ διασπασθῆναι τὸν ἄνθρωπον. ἵν' οὖν μὴ τοῦτο πάθωμεν, μηδὲ, βαρούσης τῆς πλεονεξίας κάτωθεν, ἡ ἐλεημοσύνη καταλιποῦσα ἡμᾶς ἀπέλθῃ, κουφίσωμεν ἑαυτοὺς, καὶ ἀναπετάσωμεν, ἵνα διά τε τῆς ἀπαλλαγῆς τῶν κακῶν καὶ διὰ τῆς τῶν αἰωνίων ἀγαθῶν ἐργασίας τελειωθέντες, ἐπιτύχωμεν τῶν αἰωνίων ἀγαθῶν, χάριτι καὶ φιλανθρωπίᾳ τοῦ Κυρίου ἡμῶν Ἰησοῦ Χριστοῦ, μεθ' οὗ τῷ πατρὶ, ἅμα τῷ ἁγίῳ πνεύματι, δόξα, κράτος, τιμὴ, νῦν καὶ ἀεὶ, καὶ εἰς τοὺς αἰῶνας τῶν αἰώνων. Ἀμήν.

ETONÆ,
Excusum in Collegio Regali.
M. DC. X.

Volume II continued

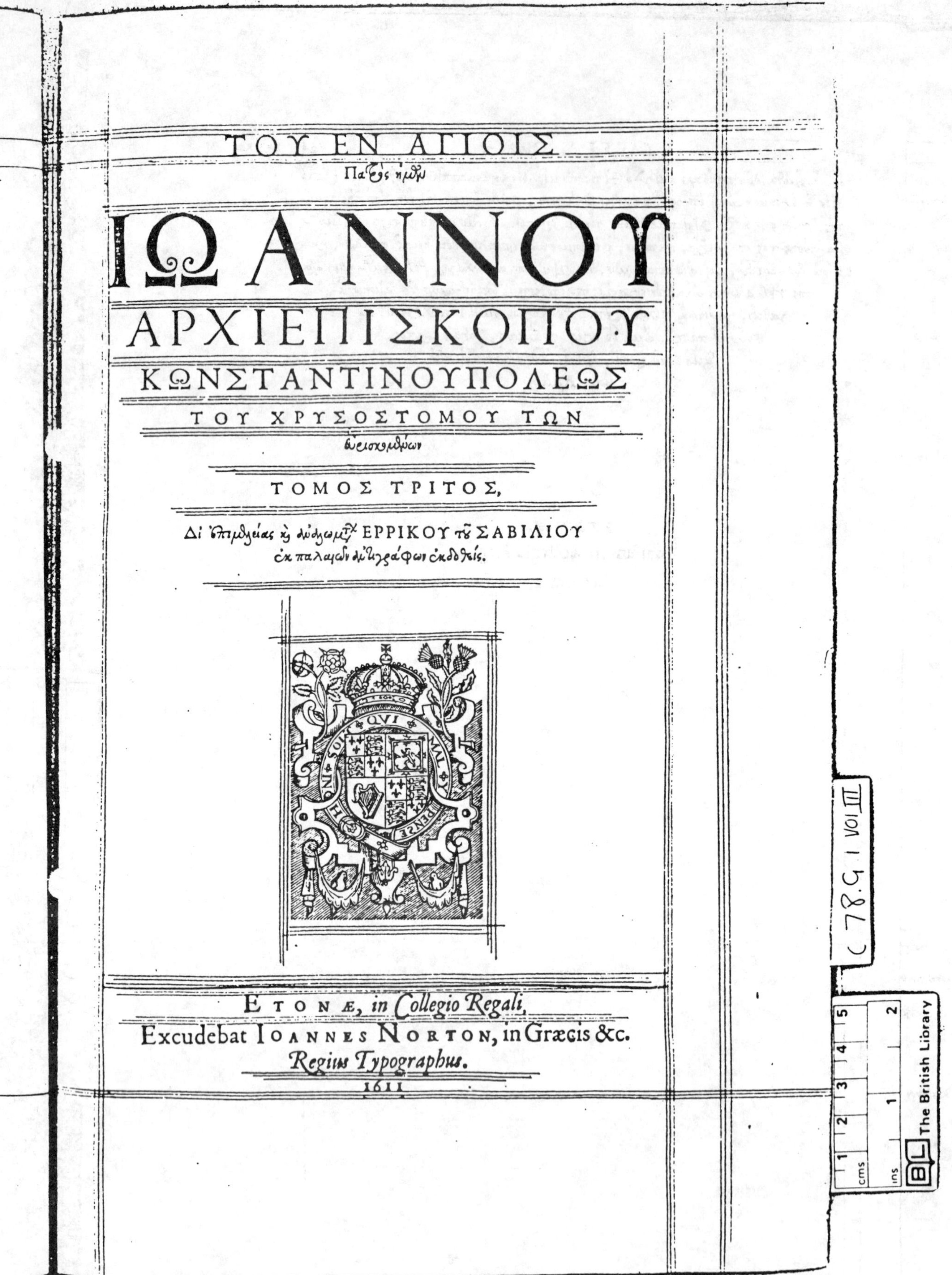
ΤΟΥ ΕΝ ΑΓΙΟΙΣ
Πατρὸς ἡμῶν

ΙΩΑΝΝΟΥ
ΑΡΧΙΕΠΙΣΚΟΠΟΥ
ΚΩΝΣΤΑΝΤΙΝΟΥΠΟΛΕΩΣ
ΤΟΥ ΧΡΥΣΟΣΤΟΜΟΥ ΤΩΝ
εὑρισκομένων

ΤΟΜΟΣ ΤΡΙΤΟΣ,

Δι' ἐπιμελείας καὶ ἀναλωμάτων ΕΡΡΙΚΟΥ τοῦ ΣΑΒΙΛΙΟΥ
ἐκ παλαιῶν ἀντιγράφων ἐκδοθείς.

ETONÆ, *in Collegio Regali*,
Excudebat IOANNES NORTON, in Græcis &c.
Regius Typographus.
1611.

Volume III or Tomos Tritos

Courtesy British Library London

παιδεύσωμεν τῆς φιλανθρωπίας, μᾶλλον δὲ καὶ πολλῷ πλείους, ἵνα μὴ μόνον σβεσθῶσιν αἱ ἁμαρτίαι, ἀλλ' ἵνα καὶ εἰς δικαιοσύνην ἡμῖν λογισθῶσιν αἱ δικαιοσύναι. οὐ γὰρ μὴ τοσαῦτα ᾖ τὰ ἀγαθά, ὡς καὶ τὰ ἐγκλήματα ἀποδύσασθαι, καὶ ἐκ τῶν ὑπολαμφθέντων εἰς δικαιοσύνην ἡμῖν λογισθῆναι, οὐδεὶς ἡμᾶς ἐξαιρήσεται τῆς κολάσεως, ἧς γένοιτο πάντας ἡμᾶς ἀπαλλαγῆναι, χάριτι καὶ φιλανθρωπίᾳ.

ETONAE,

Excusum in Collegio Regali.

M DC X.

Volume III continued

ΤΟΥ ΕΝ ΑΓΙΟΙΣ

Πατρὸς ἡμῶν

ΙΩΑΝΝΟΥ

ΑΡΧΙΕΠΙΣΚΟΠΟΥ

ΤΟΥ ΧΡΥΣΟΣΤΟΜΟΥ ΤΩΝ

εὑρισκομένων

ΤΟΜΟΣ ΤΕΤΑΡΤΟΣ,

Δι᾽ ἐπιμελείας καὶ ἀναλωμάτων ΕΡΡΙΚΟΥ τοῦ ΣΑΒΙΛΙΟΥ

ἐκ παλαιῶν ἀντιγράφων ἐκδοθέντος.

ETONÆ, *in Collegio Regali,*

Excudebat IOANNES NORTON, in Græcis &c.

Regius Typographus.

1612

Volume IV or Tomos Tetartos

Courtesy British Library London

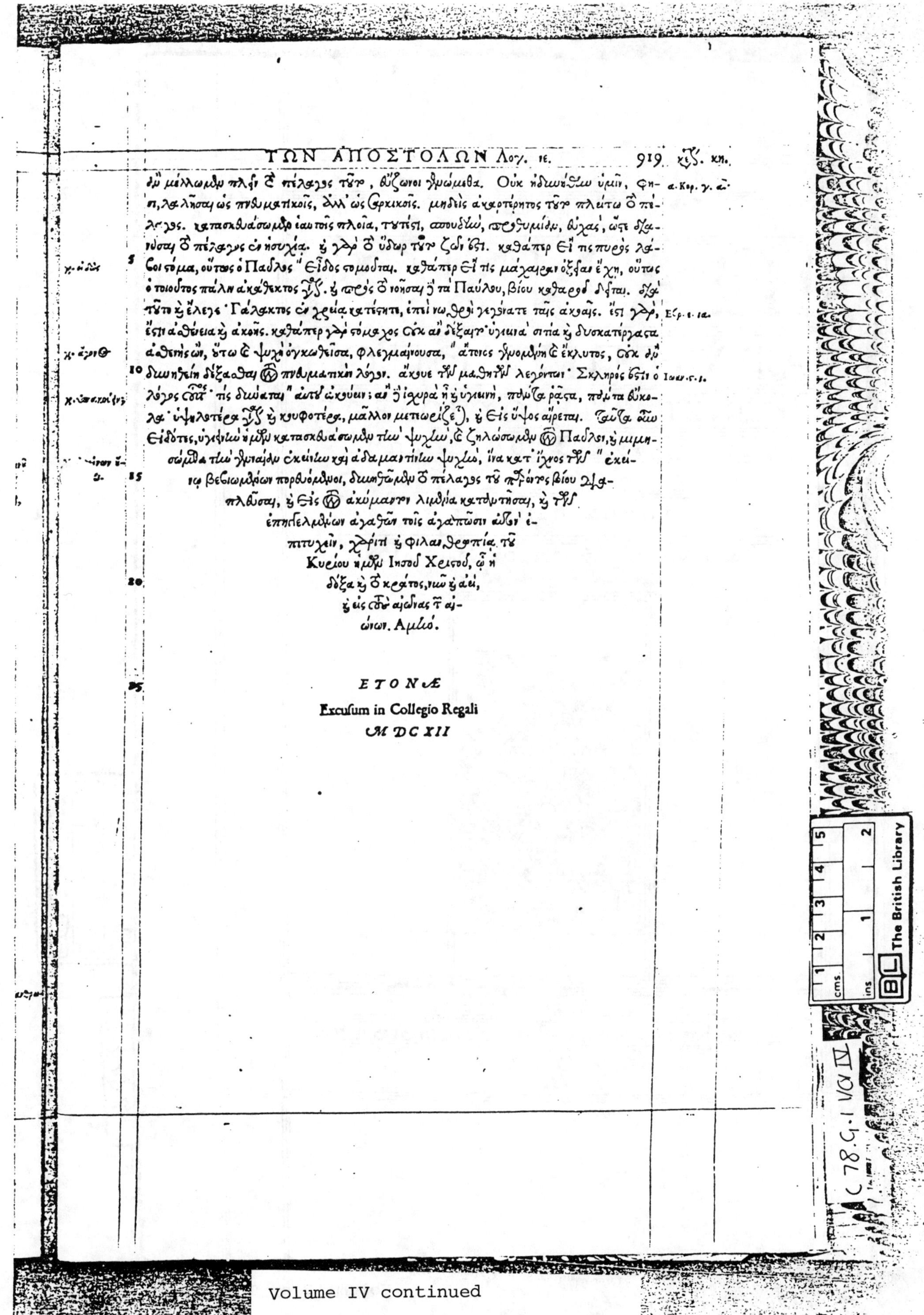

ἂν μέλλωμεν πλεῖν τὸ πέλαγος τοῦτο, εὔζωνοι γινώμεθα. Οὐκ ἠδυνήθην ὑμῖν, φησι, λαλῆσαι ὡς πνευματικοῖς, ἀλλ᾽ ὡς σαρκικοῖς. (α. Κορ. γ. α.) μηδεὶς ἀκαρτέρητος τοῦτο πλείτω τὸ πέλαγος. κατασκευάσωμεν ἑαυτοῖς πλοῖα, τουτέστι, σπουδὴν, προθυμίαν, εὐχὰς, ὥστε διαπλεῦσαι τὸ πέλαγος ἐν ἡσυχίᾳ. καὶ γὰρ τὸ ὕδωρ τοῦτο ζωή ἐστι. καθάπερ εἴ τις πυρὸς λάβοι στόμα, οὕτως ὁ Παῦλος· Εἶδος στομοῦται. καθάπερ εἴ τις μάχαιραν ὀξεῖαν ἔχῃ, οὕτως ὁ τοιοῦτος πάλιν ἀκάθεκτός ἐστι. καὶ πρὸς τὸ νοῆσαι δὲ τὰ Παύλου, βίου καθαροῦ δεῖται. διὰ τοῦτο καὶ ἔλεγε· Γάλακτος ἐν χρείᾳ καθέστηκε, ἐπεὶ νωθροὶ γεγόνατε ταῖς ἀκοαῖς. (Εβρ. ε. ιβ.) ἔστι γὰρ, ἔστιν ἀσθένεια καὶ ἀκοῆς. καθάπερ γὰρ στόμαχος οὐκ ἂν δέξαιτο ὑγιεινὰ σιτία καὶ δυσκατέργαστα ἀσθενὴς ὤν, οὕτω καὶ ψυχὴ ὀγκωθεῖσα, φλεγμαίνουσα, ἄτονος γινομένη καὶ ἔκλυτος, οὐκ ἂν δυνηθείη δέξασθαι τὸν πνευματικὸν λόγον. ἄκουε τῶν μαθητῶν λεγόντων· Σκληρός ἐστιν ὁ λόγος οὗτος· τίς δύναται αὐτοῦ ἀκούειν; (Ιωαν. ς. ξ.) ἂν δὲ ἰσχυρὰ ᾖ καὶ ὑγιεινὴ, πάντα ῥᾷστα, πάντα εὔκολα· ὑψηλοτέρα γίνεται καὶ κουφοτέρα, μᾶλλον μετεωρίζεται, καὶ εἰς ὕψος αἴρεται. Ταῦτα οὖν εἰδότες, ὑγιεινὴν ἡμῶν κατασκευάσωμεν τὴν ψυχὴν, καὶ ζηλώσωμεν τὸν Παῦλον, καὶ μιμησώμεθα τὴν γενναίαν ἐκείνην καὶ ἀδαμαντίνην ψυχὴν, ἵνα κατ᾽ ἴχνος τῶν ἐκείνῳ βεβιωμένων πορευόμενοι, δυνηθῶμεν τὸ πέλαγος τοῦ παρόντος βίου διαπλεῦσαι, καὶ εἰς τὸν ἀκύμαντον λιμένα καταντῆσαι, καὶ τῶν ἐπηγγελμένων ἀγαθῶν τοῖς ἀγαπῶσιν αὐτὸν ἐπιτυχεῖν, χάριτι καὶ φιλανθρωπίᾳ τοῦ Κυρίου ἡμῶν Ἰησοῦ Χριστοῦ, ᾧ ἡ δόξα καὶ τὸ κράτος, νῦν καὶ ἀεὶ, καὶ εἰς τοὺς αἰῶνας τῶν αἰώνων. Ἀμήν.

ETONÆ

Excusum in Collegio Regali

MDCXII

Volume IV continued

Courtesy British Library London

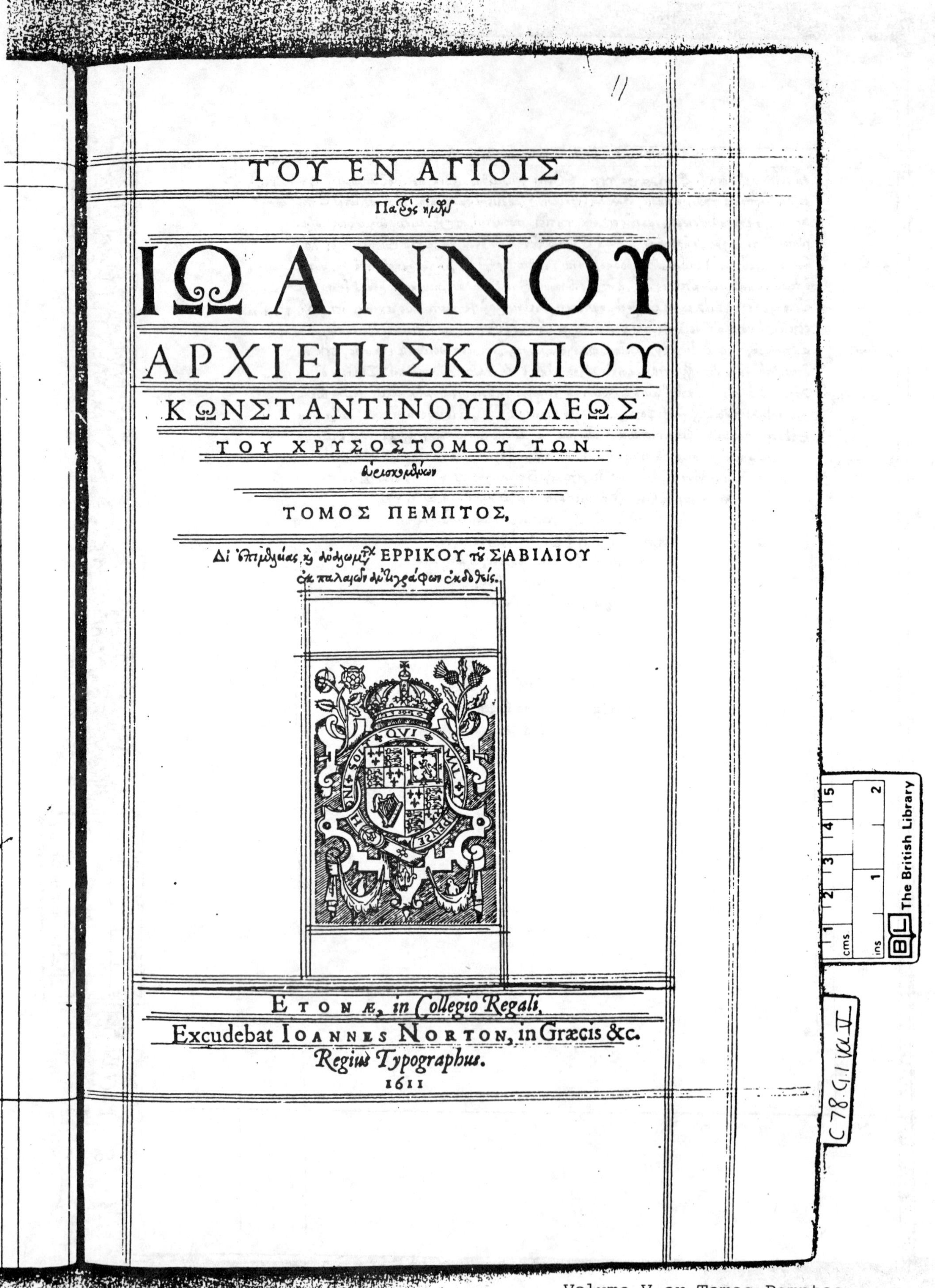

ΤΟΥ ΕΝ ΑΓΙΟΙΣ

πατρὸς ἡμῶν

ΙΩΑΝΝΟΥ

ΑΡΧΙΕΠΙΣΚΟΠΟΥ

ΚΩΝΣΤΑΝΤΙΝΟΥΠΟΛΕΩΣ

ΤΟΥ ΧΡΥΣΟΣΤΟΜΟΥ ΤΩΝ

εὑρισκομένων

ΤΟΜΟΣ ΠΕΜΠΤΟΣ,

Δι' ἐπιμελείας καὶ ἀναλωμάτων ΕΡΡΙΚΟΥ τοῦ ΣΑΒΙΛΙΟΥ

ἐκ παλαιῶν ἀντιγράφων ἐκδοθείς.

ETONÆ, *in Collegio Regali.*

Excudebat IOANNES NORTON, in Græcis &c.

Regius Typographus.

1611

Volume V or Tomos Pemptos

Courtesy British Library London

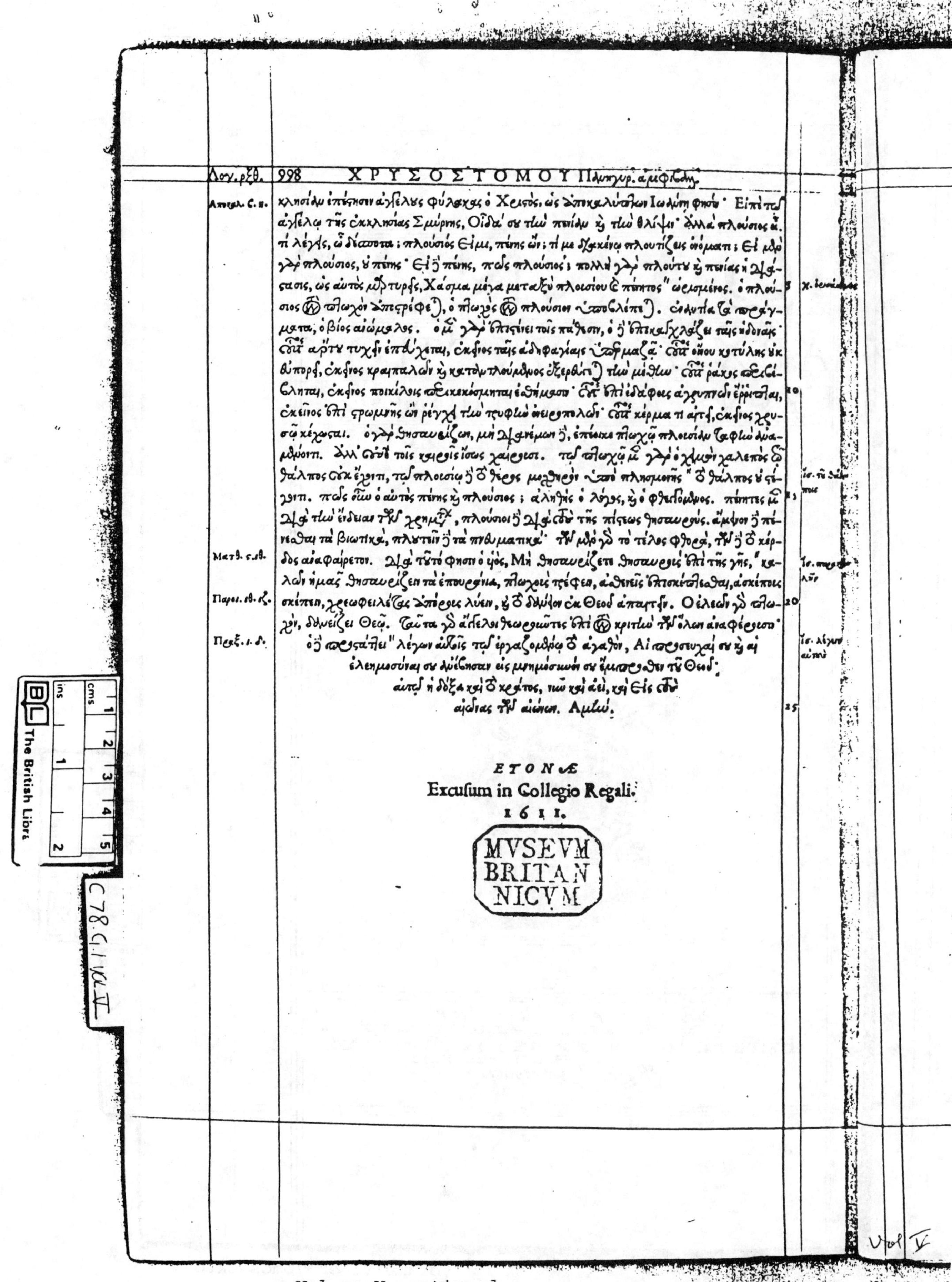

Λογ. ρξθ. 998 ΧΡΥΣΟΣΤΟΜΟΥ Πανηγυρ. ἀμφιβολ.

Ἀποκαλ. β. η. κλησίαν ἐπέστησεν ἀγγέλους φύλακας ὁ Χριστός, ὡς ἀποκαλύψεως Ἰωάννης φησίν· Εἰπὲ τῷ ἀγγέλῳ τῆς ἐκκλησίας Σμύρνης, Οἶδά σου τὴν πενίαν καὶ τὴν θλίψιν· ἀλλὰ πλούσιος εἶ. τί λέγεις, ὦ δέσποτα; πλούσιός εἰμι, πένης ὤν; τί με διακενῆς πλουτίζεις ὀνόματι; εἰ μὲν γὰρ πλούσιος, οὐ πένης· εἰ δὲ πένης, πῶς πλούσιος; πολλὴ γὰρ πλούτου καὶ πενίας ἡ διάστασις, ὡς αὐτὸς μαρτυρεῖς, Χάσμα μέγα μεταξὺ πλουσίου καὶ πένητος ὡρισμένος. ὁ πλούσιος τὸν πτωχὸν ἀποστρέφεται, ὁ πτωχὸς τὸν πλούσιον ὑποβλέπεται. ἐναντία τὰ πράγματα, ὁ βίος ἀνώμαλος. ὁ μὲν γὰρ ἐπιστένει τοῖς πάθεσι, ὁ δὲ ἐπικαγχάζει ταῖς ἡδοναῖς· οὗτος ἄρτου τυχεῖν εὔχεται, ἐκεῖνος ταῖς ἀδηφαγίαις ὑπορμάζει· οὗτος οἴνου κοτύλης οὐκ εὐπορεῖ, ἐκεῖνος κραιπαλῶν καὶ κατεμπιπλάμενος ἐξερεύγεται τὴν μέθην· οὗτος ῥάκος περιβέβληται, ἐκεῖνος ποικίλοις περικεκόσμηται ἐνδύμασιν· οὗτος ἐπὶ ἐδάφους ἀγρυπνῶν ἔρριπται, ἐκεῖνος ἐπὶ στρωμνῆς ὢν ῥέγχει τὴν τρυφὴν ὀνειροπολῶν· οὗτος κέρμα τι αἰτεῖ, ἐκεῖνος χρυσῷ κέχωσται. ὁ γὰρ θησαυρίζων, μὴ διανέμων δέ, ἔοικε πτωχῷ πλουσίαν ταφὴν ἀναμένοντι. ἀλλ' οὐδὲ τοῖς καιροῖς ἴσως χαίρουσι. τῷ πτωχῷ μὲν γὰρ ὁ χειμὼν χαλεπὸς τὸ θάλπος οὐκ ἔχοντι, τῷ πλουσίῳ δὲ τὸ θέρος μοχθηρὸν ὑπὸ πλησμονῆς τὸ θάλπος οὐ στέγοντι. πῶς οὖν ὁ αὐτὸς πένης καὶ πλούσιος; ἀληθὴς ὁ λόγος, καὶ ὁ φθεγγόμενος. πένητες μὲν διὰ τὴν ἔνδειαν τῶν χρημάτων, πλούσιοι δὲ διὰ τοὺς τῆς πίστεως θησαυρούς. ἄμεινον δὲ πένεσθαι τὰ βιωτικά, πλουτεῖν δὲ τὰ πνευματικά· τῶν μὲν γὰρ τὸ τέλος φθορά, τῶν δὲ τὸ κέρ-

Ματθ. ς. ιθ. δος ἀναφαίρετον. διὰ τοῦτό φησιν ὁ υἱός, Μὴ θησαυρίζετε θησαυροὺς ἐπὶ τῆς γῆς, καλῶν ἡμᾶς θησαυρίζειν τὰ ἐπουράνια, πτωχοὺς τρέφειν, ἀσθενεῖς ἐπισκέπτεσθαι, ἀσκέπους

Παροι. ιθ. ιζ. σκέπειν, χρεωφειλέτας ἀπόρους λύειν, καὶ τὸ δάνειον ἐκ Θεοῦ ἀπαιτεῖν. Ὁ ἐλεῶν γὰρ πτωχόν, δανείζει Θεῷ. Ταῦτα γὰρ ἄγγελοι θεωροῦντες ἐπὶ τὸ κριτήριον τῶν ὅλων ἀναφέρουσιν·

Πραξ. ι. δ. ὁ δὲ παριστάνει λέγων αὐτοῖς τῷ ἐργαζομένῳ τὸ ἀγαθόν, Αἱ προσευχαί σου καὶ αἱ ἐλεημοσύναι σου ἀνέβησαν εἰς μνημόσυνον σου ἔμπροσθεν τοῦ Θεοῦ.

αὐτῷ ἡ δόξα καὶ τὸ κράτος, νῦν καὶ ἀεί, καὶ εἰς τοὺς αἰῶνας τῶν αἰώνων. Ἀμήν.

κ. ἐστηριγμένος

ἴσ. λέγων αὐτῷ

ETONÆ
Excusum in Collegio Regali.
1611.

Volume V continued

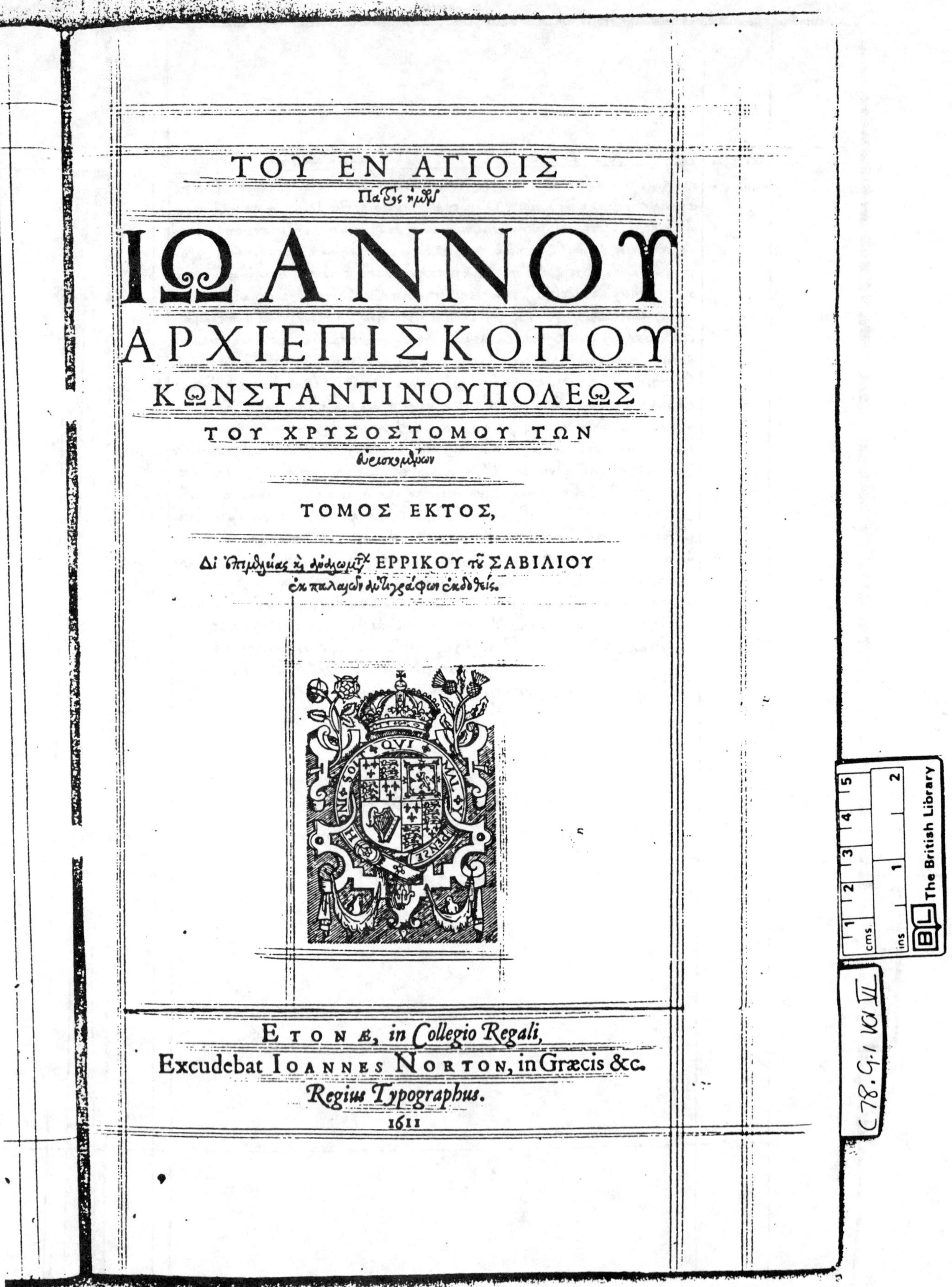

ΤΟΥ ΕΝ ΑΓΙΟΙΣ
πατρὸς ἡμῶν

ΙΩΑΝΝΟΥ
ΑΡΧΙΕΠΙΣΚΟΠΟΥ
ΚΩΝΣΤΑΝΤΙΝΟΥΠΟΛΕΩΣ
ΤΟΥ ΧΡΥΣΟΣΤΟΜΟΥ ΤΩΝ
εὑρισκομένων

ΤΟΜΟΣ ΕΚΤΟΣ,

Δι' ἐπιμελείας καὶ ἀναλωμάτων ΕΡΡΙΚΟΥ τοῦ ΣΑΒΙΛΙΟΥ
ἐκ παλαιῶν ἀντιγράφων ἐκδοθείς.

ETONÆ, *in Collegio Regali,*
Excudebat IOANNES NORTON, in Græcis &c.
Regius Typographus.
1611

Volume VI or Tomos Ektos

Courtesy British Library London

μου τὴν ἐλπίδα τῆς σωτηρίας. παρακάλεσόν μου καὶ σὺ τὴν καρδίαν, ὁ πατὴρ τῶν οἰκτιρμῶν καὶ Θεὸς πάσης παρακλήσεως· ψυχαγώγησον καὶ ἀγάθυνον, ἡ πηγὴ τοῦ ἐλέους καὶ τῶν ἀγαθῶν. πολλὰ ἐποίησας ἀπὸ τοῦ αἰῶνος, μεγάλα καὶ θαυμαστά, ἔνδοξά τε καὶ ἐξαίσια, ὧν οὐκ ἔστιν ἀριθμός. ἀλλ' εἰ ἐμὲ τὸν ἄσωτον σώσεις, εἰ ἐμὲ τὸν ἀνάξιον παραστήσεις, πλείω καὶ μείζω θαυμαστωθήσῃ· ὅτι τοσαύτη σου τῆς ἐλεημοσύνης καὶ τῆς φιλανθρωπίας ἡ δύναμις, ὥστε καὶ ἀπὸ βορβόρου μαργαρίτην ποιεῖν, καὶ ἀπὸ τοῦ εἶναι υἱὸν γεέννης, υἱὸν βασιλείας ἀποτελεῖν. καὶ ἔτι κεκράξομαι πρὸς τὸν Κύριόν μου, καὶ πρὸς τὸν Θεόν μου δεηθήσομαι· κυβέρνησον τὸ ὑπόλοιπόν μου τῆς ζωῆς κατὰ τὸ θέλημά σου· στήριξόν με εἰς τὸν φόβον σου· στερέωσόν με ἐν τῇ ἀγάπῃ σου, καὶ ἐν τῷ πλήθει τῆς χρηστότητός σου χρηστόν μοι τέλος καὶ ἄξιον τῆς φιλανθρωπίας σου δώρησαι, καὶ τὰ ὀστᾶ καὶ τὰς ἁρμονίας μου καὶ πᾶσαν τὴν σύνθεσιν τῆς ὑποστάσεώς μου ἐν ἐλέει καὶ φιλανθρωπίᾳ σου ἔπιδε, καὶ εἰς ἀνέσεως καὶ εἰς ἀναπαύσεως τόπον τὴν ἀθλίαν μου ψυχὴν ὁδηγήσας, ἀποκατάστησον, ὅτι πολλαὶ μοναὶ παρὰ σοὶ ἑκάστῳ κατ' ἀξίαν διανεμόμεναι. ἔτι δέομαι καὶ παρακαλῶ· δός δή, Κύριε, καὶ χάριν συνέσεως τῇ ἐμῇ ἀναξιότητι τοῦ διαγινώσκεσθαι τὰ σοὶ εὐάρεστα, κἀμοὶ συμφέροντα· καὶ μὴ μόνον διαγινώσκειν, ἀλλὰ καὶ διαπράττεσθαι· τὸ μὴ συναρπάζεσθαι καὶ συναπάγεσθαι τοῖς ματαίοις, τὸ μὴ διαπράττεσθαι τὰ μὴ δέοντα, τὸ ἀκενοδόξως ταπεινοῦσθαι τοῖς ταπεινοῖς, καὶ τοῖς πάσχουσι συμπαθεῖν, καὶ τοῖς ἁμαρτάνουσι συγχωρεῖν. οἶδα γὰρ ὡς, εἰ μὴ ἀφήσω, οὐκ ἀφεθήσομαι. διὰ τοῦτο, παρακαλῶ, συγχώρησον πάντα πᾶσι τοῖς ἁμαρτάνουσιν εἰς ἐμέ. οὐ γάρ εἰσιν οὗτοι αἴτιοι, ἀλλ' ἐγὼ ὁ ἄθλιος, ὁ μὴ ποιῶν τὸ θέλημά σου, καὶ μὴ φυλάττων τὰ προστάγματά σου. τοὺς δὲ ἀγαπῶντας ἡμᾶς ἀντάμειψαι ταῖς πλουσίαις σου δωρεαῖς· τὸν δὲ πνευματικόν μου πατέρα, καὶ τοὺς ἀδελφούς, οὓς σὺ δέδωκας, εὔσπλαγχνε, κρίμασιν οἷς ἐπίστασαι σὺν ἐμοὶ φιλανθρώπως οἰκτειρήσας, ἐλέησον. ταῦτα δέ μου τῆς προσευχῆς τὰ ῥήματα ἔστωσαν ὑπερεντυγχάνοντά μου καὶ ζῶντος καὶ θανόντος· αὕτη ἡ ἐξομολόγησις καὶ τὰ δάκρυα ὡς θυμίαμα ἐνώπιόν σου κατευθυνθήτωσαν· ἐγὼ δὲ καθ' ἑκάστην ἀναμένω τοῦ θανάτου τὸ ἀπαραίτητον. καὶ τὸ μὲν σῶμά μου τὸ ἄθλιον ταφῇ παραδοθὲν διαφθαρήσεται, καὶ εἰς τὰ, ἐξ ὧν συνετέθη, ἀναλυθήσεται, ὅπερ ἀναστήσεις ὁ ζωοδότης ἄφθαρτον ἐν τῷ τῆς παλιγγενεσίας καιρῷ· τὸ δὲ πνεῦμά μου εἰς χεῖράς σου παρατίθημι. ἀνάπαυσον, ἅγιε δέσποτα, ἐν φωτὶ ζώντων, καὶ ἐν τῇ κατοικίᾳ τῶν εὐφραινομένων, καὶ τοὺς ἐμοὺς γεννήτορας προγόνους, καὶ ἀδελφούς, σοὺς δὲ οἰκέτας εὐγνώμονας· ὅτι εἰ καὶ ἡμάρτομεν, ἀλλ' οὐκ ἀπέστημεν ἀπὸ σοῦ, οὐδὲ διεπετάσαμεν χεῖρας ἡμῶν πρὸς Θεὸν ἀλλότριον· ἀλλὰ σὲ ἔγνωμεν, καὶ σὲ ἠγαπήκαμεν, καὶ σοὶ πεπιστεύκαμεν, καὶ σὲ προσκυνοῦμεν τὸν ἕνα ἐν τριάδι Θεόν. ἐν σοί τε προσευχόμεθα, καὶ ἐν σοὶ τὰς τῆς σωτηρίας ἀναρτῶμεν ἐλπίδας. ἐλέησον ἡμᾶς κατὰ τὸ μέγα σου ἔλεος, καὶ σῶσον εἰς τὴν ἐπουράνιον καὶ αἰώνιόν σου βασιλείαν. ναὶ δή, Κύριέ μου, Κύριε, οὕτω γενέσθω ταῦτα ἐν ἡμῖν τοῖς ἐλπίζουσιν ἐπὶ σὲ διὰ τὴν πολλήν σου καὶ ἀνυπέρβλητον ἀγαθότητα, καὶ διὰ τὴν ἄφατόν σου εὐσπλαγχνίαν καὶ φιλανθρωπίαν, πρεσβείαις τῆς πανενδόξου, πανυμνήτου, ὑπερευλογημένης, καὶ κεχαριτωμένης δεσποίνης ἡμῶν, ὑπεραγίας θεοτόκου, καὶ ἀειπαρθένου Μαρίας, τῶν ἐπουρανίων καὶ νοερῶν δυνάμεων, καὶ πάντων τῶν ἀπ' αἰῶνός σοι εὐαρεστησάντων.

Ἀμήν.

ETONÆ
Excusum in Collegio Regali
M DC XI.

Volume VI continued

ΤΟΥ ΕΝ ΑΓΙΟΙΣ

πατρὸς ἡμῶν

ΙΩΑΝΝΟΥ

ΑΡΧΙΕΠΙΣΚΟΠΟΥ

ΚΩΝΣΤΑΝΤΙΝΟΥΠΟΛΕΩΣ

ΤΟΥ ΧΡΥΣΟΣΤΟΜΟΥ ΤΩΝ

εὑρισκομένων

ΤΟΜΟΣ ΕΒΔΟΜΟΣ,

Δι' ἐπιμελείας καὶ ἀναλωμάτων ΕΡΡΙΚΟΥ τοῦ ΣΑΒΙΛΙΟΥ

ἐκ παλαιῶν ἀντιγράφων ἐκδοθείς.

ETONÆ, *in Collegio Regali.*

Excudebat IOANNES NORTON, in Græcis &c.

Regius Typographus.

1612

Volume VII or Tomos Hebdomos

Courtesy British Library London

Sermo ipsius Severiani de pace, cum susceptus esset à beato Iohanne episcopo Constantinopolitano.

IN adventu Domini & Salvatoris nostri atq; in præsentia eius corporali, angeli ducentes choros cœlestes evangelizabant pastoribus, dicentes: Annuntiamus vobis hodie gaudium magnum, quod erit omni populo. Ab ipsis enim sanctis angelis etiam mutuatam nos vocem, annuntiamus vobis hodie gaudium magnum. *Luc. 2. 10.* Hodie enim in pace Ecclesia est, & hæretici in ira. Hodie Ecclesiæ navis in portu est, & hęreticorum furor jactatur in fluctibus. Hodie Pastores Ecclesiæ in securitate, & hæretici in perturbatione sunt. Hodie oves Domini in tuto, & lupi insaniunt. Hodie vinea Domini in abundantia, & operarii iniquitatis in egentia. Hodie populus Christi exaltatus est, & inimici veritatis humiliati sunt. Hodie Christus in lætitia, & diabolus in luctu. Hodie Angeli in exultatione, & dæmones in confusione. Et quid opus est multa dicere? Hodie Christus, qui est Rex Pacis, cum sua pace procedens, fugavit omne dissidium, dissensiones depulit, discordiam perturbavit. Et sicut cœlum splendor solis, ita ecclesiam fulgor pacis illuminat. O quàm desiderabile nomen Pacis, quàm religionis Christianæ stabile fundamentum, & altaris dominici cœlestis armatura. Et quid possumus de Pace proloqui? Pax nomen est ipsius Christi, sicut dicit & Apostolus: Quia Christus pax nostra, qui fecit utráq; unum, quæ nequaquam fide, sed invidia diaboli dissidebant. *Eph. 2. 14.* Verum sicut procedente rege & plateæ mundantur, & tota civitas diversis floribus & ornatibus coronatur, ut nihil sit quod minus dignũ vultui regis appareat: ita & nunc procedente Christo rege Pacis, omne quod triste est, auferatur è medio, & illucescente veritate fugetur mendacium: fugiat discordia, resplendescente concordia. Et sicut frequenter fieri vidimus, ubi regum vel fratrum tabulæ pinguntur, ut in utrisq; unanimitatis declarentur insignia, artifex picto fœmineo habitu post tergum utriúsq; concordiam statuit, brachiis suis utrúmq; complectentem, indicans quòd hi qui videntur corporibus separati, sententiis & voluntate conveniant: ita nunc pax Domini media assistens, & utrúmq; nostrûm gremio palpante connectens, discreta corpora in unum convenire animum, ulnis jungentibus docet. In qua sine dubio completur sermo propheticus, qui ait: Et erit consilium pacificum inter utrósque. Et hesterno quidem die pater noster communis, evangelico pacem sermone præfatus est: hodie verò nos verba pacis expotiamus. Ipse nos heri resupinis manibus in verbo pacis excepit, & nos hodie dilatato pectore, ulnisq; patentibus ad dominum cum muneribus pacis occurramus. Jam bella destructa sunt, pulchritudo pacis obtinuit. Nunc in luctu est diabolus, & in lamentatione omnis dæmonum turma, nunc in cœlestibus lætitia, & in angelis exultatio, quibus specialis familiaris pax. Hoc enim etiam cœlestes admirantur virtutes opus, quas fons ejus habet perennis, ex quo etiam terrena hæc nostra guttis quibusdam exinde stillantibus irrorantur. Et ideo etiamsi in terris pax, laudis ejus splendor redundat in cœlum, laudant eam cœlestes angeli, & dicunt: Gloria in excelsis Deo, & in terra pax hominibus bonæ voluntatis. Vides quomodo cœlestes omnes & terrestres invicem sibi munera pacis emittunt? *Luc. 2. 14.* Cœlestes angeli pacem terris annuntiant, sancti in terris Christum, qui est pax nostra, collaudant in cœlestibus positum, & mysticis choris acclamant, Osanna in excelsis. Dicamus ergo & nos, Gloria in altissimis Deo, qui humiliavit diabolum, & exaltavit Christum suum. Gloria in altissimis Deo, qui discordiam fugat, & pacem statuit. *Matt. 21. 9.* Dico enim vobis artem diaboli, cujus neq; vos astutiam ignoratis. Vidit satanas firmitatem fidei, stabilitatem in ea pietate dogmatum septam, vidit & operum bonorum fructibus abundantem: & ideo pro his omnibus ad insaniam venit, & rabie furoris exarsit, ut scinderet amicitiam & evelleret charitatem, ut disrumperet pacem: sed pax Domini semper sit nobiscum, in Christo Iesu Domino nostro, cum quo est Deo Patri & Spiritui sancto gloria, in secula seculorum. Amen.

ETONÆ, *Excusum in Collegio Regali.*
1612.

Volume VII continued

Courtesy British Library London

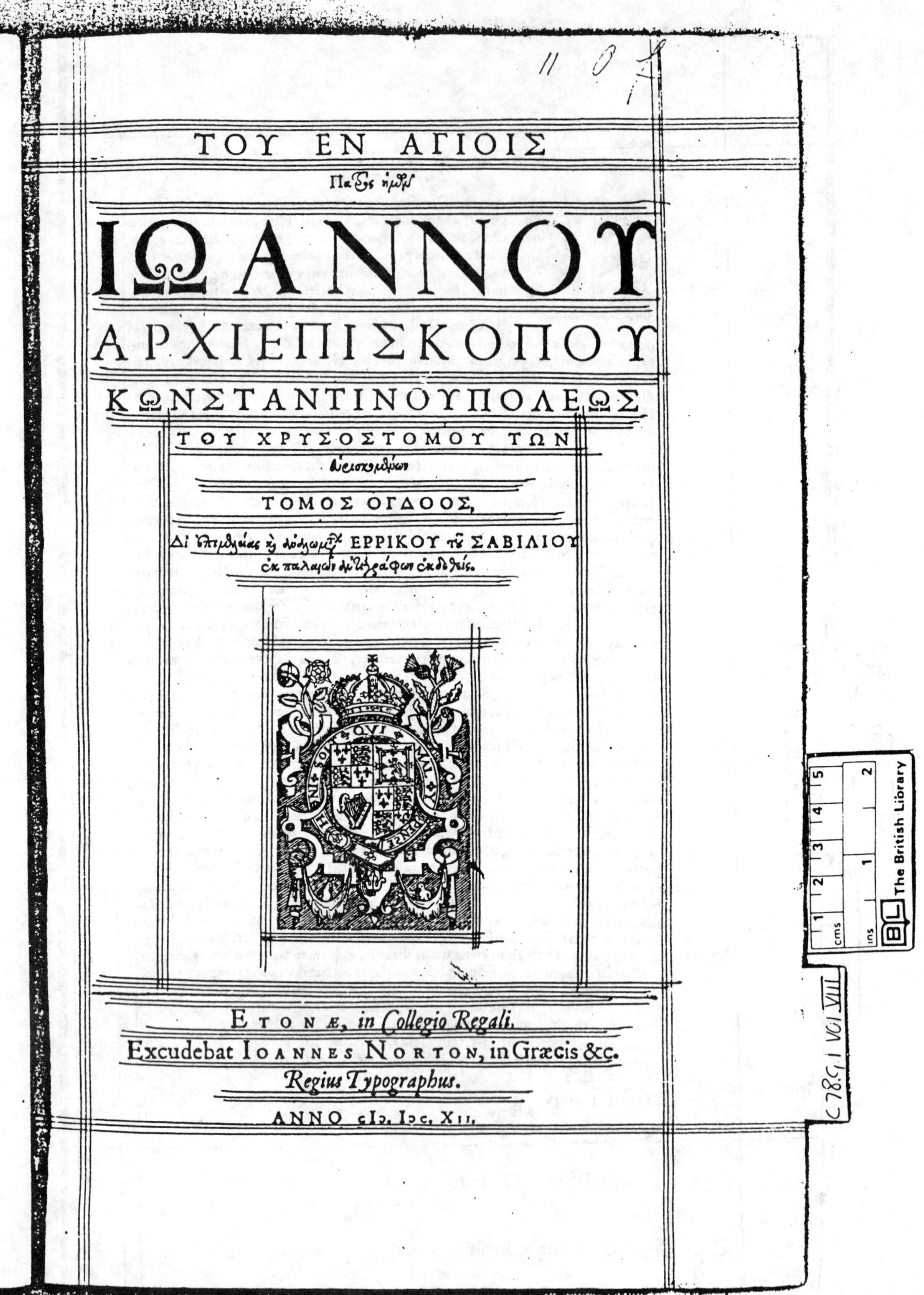

ΤΟΥ ΕΝ ΑΓΙΟΙΣ

πατρὸς ἡμῶν

ΙΩΑΝΝΟΥ

ΑΡΧΙΕΠΙΣΚΟΠΟΥ

ΚΩΝΣΤΑΝΤΙΝΟΥΠΟΛΕΩΣ

ΤΟΥ ΧΡΥΣΟΣΤΟΜΟΥ ΤΩΝ

εὑρισκομένων

ΤΟΜΟΣ ΟΓΔΟΟΣ,

Δι' ἐπιμελείας καὶ ἀναλωμάτων ΕΡΡΙΚΟΥ τοῦ ΣΑΒΙΛΙΟΥ

ἐκ παλαιῶν ἀντιγράφων ἐκδοθείς.

ETONÆ, in Collegio Regali.

Excudebat IOANNES NORTON, in Græcis &c.

Regius Typographus.

ANNO cIↃ. IↃc. XII.

Volume VIII or Tomos Ocdoos

Courtesy British Library London

μδ'. Πρὸς τοὺς ἀκροατάς, ὥστε προσέχειν τοῖς λεγομένοις, καὶ προκόπτειν ἐν τῇ κατὰ Θεὸν ἀγάπῃ.

με'. Προτροπὴ περὶ φιλοξενίας, καὶ ὅτι ὁ τοὺς ἐλαχίστους τῶν ἀδελφῶν δεχόμενος διὰ τὸν Χριστὸν, πλείονα μισθὸν ἔχει.

μς'. Ὅτι οὐ δεῖ ἀλγεῖν διὰ τὸ εἶναι αἱρέσεις, καὶ περὶ τῶν σκανδάλων.

μζ'. Ὅτι δεῖ ἡμᾶς μάρτυρας εἶναι Χριστοῦ διὰ ὀρθότητος βίου.

μη'. Περὶ ἐπιεικείας καὶ ἀνδρείας, καὶ ὅτι ἀντίκειται τούτοις θρασύτης καὶ ἀνανδρία.

μθ'. Ὅτι δι' ὧν δοκοῦμεν βλάπτεσθαι, διὰ τούτων ὠφελούμεθα· τὸ ὑπόδειγμα ἐκ τοῦ Ἰωσήφ, καὶ Ἰακώβ, καὶ ὅτι οὐ δεῖ γυναῖκας πλουσίας ἄγεσθαι.

ν'. Περὶ τοῦ μὴ δεῖν ἀνθυβρίζειν, ἀλλὰ φέρειν τοὺς ὑβρίζοντας, καὶ προσδέχεσθαι ἀσμένως τὰς καταλλαγάς.

να'. Ὅτι οὐδεὶς ἡμᾶς βλάψαι δύναται, ἐὰν μὴ ἑαυτοὺς βλάψωμεν.

νβ'. Ὅτι ὁ κρατηθεὶς τῷ θείῳ ἔρωτι, οὐδενὸς τῶν ἐπὶ γῆς λόγον ποιεῖται.

νγ'. Ὅτι χρήσιμον τὸ ἔχειν ἐν οἰκίᾳ ἅγιον ἄνδρα, καὶ πείθεσθαι τούτῳ, καὶ ὅτι δεῖ καὶ νῦν ὥσπερ πρότερον πείθεσθαι τῷ Παύλῳ συμβουλεύοντι, πρὶν ἐν πείρᾳ γενώμεθα τῶν κακῶν.

νδ'. Ὅτι χρήσιμοι εἰσὶ πειρασμοί, καὶ ὅτι οὐκ ἀεὶ ὁ δαίμων ἡμῖν τῆς κακίας αἴτιος ἐστί.

νε'. Εἰς Παῦλον τὸν ἀπόστολον.

Volume VIII continued

STC 14629a Second or Composite Edition

Series Title Page: 1613

Courtesy York Minster Library

ΤΟΥ ΕΝ ΑΓΙΟΙΣ

πατρὸς ἡμῶν

ΙΩΑΝΝΟΥ

ΑΡΧΙΕΠΙΣΚΟΠΟΥ

ΚΩΝΣΤΑΝΤΙΝΟΥΠΟΛΕΩΣ

ΤΟΥ ΧΡΥΣΟΣΤΟΜΟΥ ΤΩΝ

εὑρισκομένων

ΤΟΜΟΣ ΠΡΩΤΟΣ,

Δι' ἐπιμελείας καὶ ἀναλωμάτων ΕΡΡΙΚΟΥ τοῦ ΣΑΒΙΛΙΟΥ
ἐκ παλαιῶν ἀντιγράφων ἐκδοθείς.

ETONÆ, *in Collegio Regali.*
Excudebat IOANNES NORTON, in Græcis &c.
Regius Typographus.

ANNO cIↄ. Iↄc. XII.

STC 14629a

Volume I or Tomos Protos

Courtesy York Minster Library

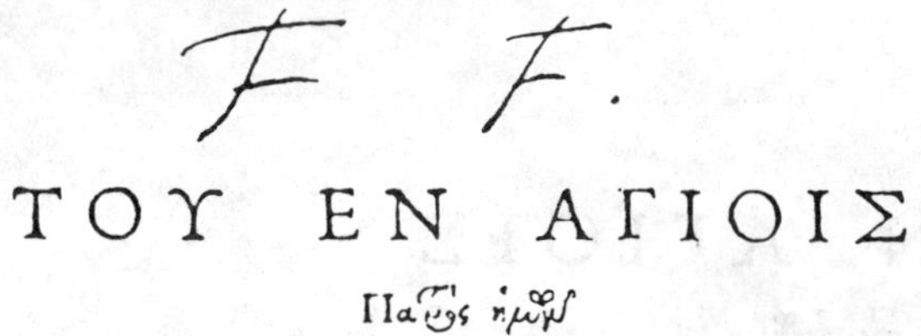

ΤΟΥ ΕΝ ΑΓΙΟΙΣ
Πατρὸς ἡμῶν

ΙΩΑΝΝΟΥ
ΑΡΧΙΕΠΙΣΚΟΠΟΥ
ΚΩΝΣΤΑΝΤΙΝΟΥΠΟΛΕΩΣ
ΤΟΥ ΧΡΥΣΟΣΤΟΜΟΥ ΤΩΝ
εὑρισκομένων

ΤΟΜΟΣ ΔΕΥΤΕΡΟΣ,

Δι' ἐπιμελείας καὶ ἀναλωμάτων ΕΡΡΙΚΟΥ τοῦ ΣΑΒΙΛΙΟΥ
ἐκ παλαιῶν ἀντιγράφων ἐκδοθείς.

ETONÆ, *in Collegio Regali,*
Excudebat IOANNES NORTON, in Græcis &c.
Regius Typographus.

ANNO cIↃ. IↃc. XII.

STC 14629a

Volume II or Tomos Deuteros

ΤΟΥ ΕΝ ΑΓΙΟΙΣ

πατρὸς ἡμῶν

ΙΩΑΝΝΟΥ

ΑΡΧΙΕΠΙΣΚΟΠΟΥ

ΚΩΝΣΤΑΝΤΙΝΟΥΠΟΛΕΩΣ

ΤΟΥ ΧΡΥΣΟΣΤΟΜΟΥ ΤΩΝ

εὑρισκομένων

ΤΟΜΟΣ ΤΡΙΤΟΣ,

Δι' ἐπιμελείας καὶ ἀναλωμάτων ΕΡΡΙΚΟΥ τοῦ ΣΑΒΙΛΙΟΥ
ἐκ παλαιῶν ἀντιγράφων ἐκδοθείς.

E T O N Æ, *in Collegio Regali.*
Excudebat I O A N N E S N O R T O N, in Græcis &c.
Regius Typographus.

ANNO cIↄ. Iↄc. XII.

STC 14629a

Volume III or Tomos Tritos

Courtesy York Minster Library

ΤΟΥ ΕΝ ΑΓΙΟΙΣ

Πατρὸς ἡμῶν

ΙΩΑΝΝΟΥ ΑΡΧΙΕΠΙΣΚΟΠΟΥ ΚΩΝΣΤΑΝΤΙΝΟΥΠΟΛΕΩΣ

ΤΟΥ ΧΡΥΣΟΣΤΟΜΟΥ ΤΩΝ

εὑρισκομένων

ΤΟΜΟΣ ΤΕΤΑΡΤΟΣ,

Δι' ἐπιμελείας καὶ ἀναλωμάτων ΕΡΡΙΚΟΥ τοῦ ΣΑΒΙΛΙΟΥ
ἐκ παλαιῶν ἀντιγράφων ἐκδοθείς.

ETONÆ, *in Collegio Regali.*
Excudebat IOANNES NORTON, in Græcis &c.
Regius Typographus.
ANNO CIↃ. IↃC. XII.

STC 14629a

Volume IV or Tomos Tetartos

ΤΟΥ ΕΝ ΑΓΙΟΙΣ

Πατρὸς ἡμῶν

ΙΩΑΝΝΟΥ

ΑΡΧΙΕΠΙΣΚΟΠΟΥ

ΚΩΝΣΤΑΝΤΙΝΟΥΠΟΛΕΩΣ

ΤΟΥ ΧΡΥΣΟΣΤΟΜΟΥ ΤΩΝ

εὑρισκομένων

ΤΟΜΟΣ ΠΕΜΠΤΟΣ,

Δι' ἐπιμελείας καὶ ἀναλωμάτων ΕΡΡΙΚΟΥ τοῦ ΣΑΒΙΛΙΟΥ

ἐκ παλαιῶν ἀντιγράφων ἐκδοθείς.

ETONÆ, *in Collegio Regali*.

Excudebat IOANNES NORTON, in Græcis &c.

Regius Typographus.

ANNO cIↃ. IↃc. XII.

STC 14629a

Volume V or Tomos Pemptos

ΤΟΥ ΕΝ ΑΓΙΟΙΣ

Πατρὸς ἡμῶν

ΙΩΑΝΝΟΥ

ΑΡΧΙΕΠΙΣΚΟΠΟΥ

ΚΩΝΣΤΑΝΤΙΝΟΥΠΟΛΕΩΣ

ΤΟΥ ΧΡΥΣΟΣΤΟΜΟΥ ΤΩΝ

εὑρισκομένων

ΤΟΜΟΣ ΕΚΤΟΣ,

Δι' ἐπιμελείας καὶ ἀναλωμάτων ΕΡΡΙΚΟΥ τοῦ ΣΑΒΙΛΙΟΥ

ἐκ παλαιῶν ἀντιγράφων ἐκδοθείς.

E T O N Æ, *in Collegio Regali.*

Excudebat I O A N N E S N O R T O N, in Græcis &c.

Regius Typographus.

A N N O cIↃ. IↃc. XII.

STC 14629a

Volume VI or Tomos Ektos

ΤΟΥ ΕΝ ΑΓΙΟΙΣ
Πατρὸς ἡμῶν

ΙΩΑΝΝΟΥ
ΑΡΧΙΕΠΙΣΚΟΠΟΥ
ΚΩΝΣΤΑΝΤΙΝΟΥΠΟΛΕΩΣ

ΤΟΥ ΧΡΥΣΟΣΤΟΜΟΥ ΤΩΝ
εὑρισκομένων

ΤΟΜΟΣ ΕΒΔΟΜΟΣ,

Δι' ἐπιμελείας καὶ ἀναλωμάτων ΕΡΡΙΚΟΥ τοῦ ΣΑΒΙΛΙΟΥ
ἐκ παλαιῶν ἀντιγράφων ἐκδοθείς.

E T O N Æ, *in Collegio Regali.*
Excudebat I O A N N E S N O R T O N, in Græcis &c.
Regius Typographus.
ANNO cIↃ. IↃC. XII.

STC 14629a

Volume VII or Tomos Hebdomos

F F

ΤΟΥ ΕΝ ΑΓΙΟΙΣ
πατρὸς ἡμῶν
ΙΩΑΝΝΟΥ
ΑΡΧΙΕΠΙΣΚΟΠΟΥ
ΚΩΝΣΤΑΝΤΙΝΟΥΠΟΛΕΩΣ
ΤΟΥ ΧΡΥΣΟΣΤΟΜΟΥ ΤΩΝ
εὑρισκομένων

ΤΟΜΟΣ ΟΓΔΟΟΣ,

Δι' ἐπιμελείας ἢ ἀναλωμάτων ΕΡΡΙΚΟΥ τοῦ ΣΑΒΙΛΙΟΥ
ἐκ παλαιῶν ἀντιγράφων ἐκδοθείς.

ETONÆ, in Collegio Regali.
Excudebat IOANNES NORTON, in Græcis &c.
Regius Typographus.
ANNO cIↃ. IↃc. XII.

STC 14629a

Volume VIII or Tomos Ocdoos

Sancti

GREGORII NAZIANZENI IN IVLIANVM INVECTIVÆ DVÆ.

Cum Scholiis Græcis nunc primùm editis, & ejusdem Authoris nonnullis aliis quorum syllabum sequens Pag. continet.

Omnia, ex Bibliothecâ Clarissimi viri D HENRICI SAVILII, Edidit *R. Montagu.*

ETONAE, *in Collegio Regali,*
Excudebat *Ioannes Norton*, in Græcis, &c.
Regius Typographus.
1610.

Isaacus Casaubonus

STC 12346

ΙΩΑΝΝΟΥ ΜΕΜΦΙΤΟΥ ΕΙΣ ΒΙΒΛΟΝ
τῶν μζ. λόγων τοῦ Θεολόγου.

Γρηγόριος, μερόπεσσι, θεηγόρος ἄλθεα μύθων,
Τεσσαράκοντα καὶ ἑπτὰ σοφοῖς βίβλοισι χαράξας,
Αἱρετικῆς ἀπέκερσε θεομάχον ἄλγος Ἐνυοῦς,
Δήμιον αὐδήεντι νόου θώρηκι πυκάσσας·
Καί γε μελυσπαλέος ἀπατηλόεργος αὐχένας ὕδρης
Ἰὸν ὀλεσσιπόνοιο διαβλύζοντας ἰωῆς·
Εὖτε γὰρ ἑπταλόφοιο θεηδόχα πώεα Ῥώμης
Δόγμασι μιστύλλοντο πολυσχιδέεσσι νομήων,
Λυγρῇ ἀσημάντοισι ῥοπαῖς νόμον ἀμφιχέοντα,
Μοῦνος Χριστὸς ἔην ἐπιτάρροθος. ἀθεσφάτως γάρ,
Πλαζομένης προσαπίσσεσι φίλης ἰθύντορα ποίμνης,
Γρηγόριον ξυνάεθλον ἀμαιμακέτου Βασιλείου,
Καππαδοκῶν μετὰ γαῖαν ἐς ὁπλοτέρης χθόνα Ῥώμης
Εἴρυσεν αὐτολίπητον ἑωσφόρος. οὐκ ἔτι οἱ ἄκ[illegible]ις
Ἡδυεπὴς στομάτων ἀνεπάλλετο· πασσυδίῃ δὲ
Οἶδμα κελαινοπτέροιο διαυγάζουσα μελιοπῆς,
Θυμοπαγὴς καὶ βαιὸν αἰσχλίζεσκεν ὁμίχλην

ERRATA.

Pag. 4. lin. 13. [illegible] lege [illegible]. 9. 25. [illegible] lege [illegible]. 13. distingue in scb. sic τέλος καὶ τὸ τέλεσμα, Κατὰ [illegible] ἐν τελέεσσι. 16. in scb. [illegible] lege [illegible]. 18. 20. sit comma non colon post Καίσαρι. 19. 17. deest colon post [illegible]. 23. 16. Rectius opinor, per interrogat. 29. scb. [illegible] lege [illegible]. 52. 9. αἱρούμενος lege [illegible]. 56. 6. [illegible] lege [illegible]. 59. 8. [illegible] lege [illegible]. 77. 9. sit plena distinctio post [illegible] in scb. ἐξ ὄντος lege ἐξόντος. 89. in scb. [illegible] lege [illegible]. 92. in schol. [illegible] lege ἀπώλοντο. Ibid. [illegible] lege [illegible]. 102. 2. [illegible] lege Θεὸς. 103. 1. sit comma post [illegible]. 110. scb. [illegible] lege [illegible].

Sancti Gregorii - STC 12346 continued

III

S E C R E T

PROVINCIAL

P R I N T I N G

(SEE MAP OVERLEAF)

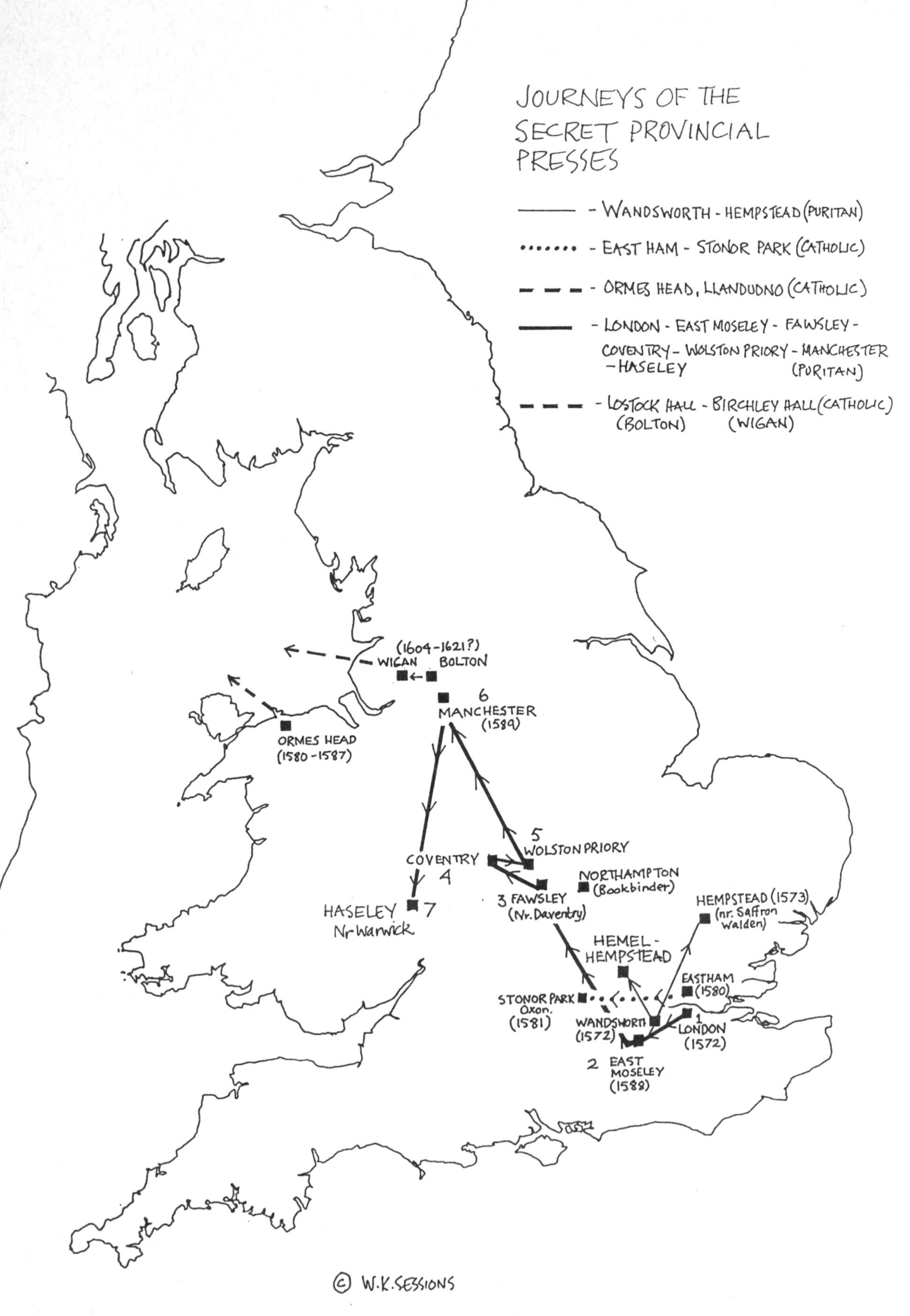

JOURNEYS OF THE SECRET PROVINCIAL PRESSES
- WANDSWORTH - HEMPSTEAD (PURITAN)
- EAST HAM - STONOR PARK (CATHOLIC)
- ORMES HEAD, LLANDUDNO (CATHOLIC)
- LONDON - EAST MOSELEY - FAWSLEY - COVENTRY - WOLSTON PRIORY - MANCHESTER - HASELEY (PURITAN)
- LOSTOCK HALL (BOLTON) - BIRCHLEY HALL (WIGAN) (CATHOLIC)
(1604-1621?)
WIGAN
BOLTON
6
MANCHESTER
(1589)
ORMES HEAD
(1580-1587)
5
WOLSTON PRIORY
COVENTRY
4
NORTHAMPTON
(Bookbinder)
3 FAWSLEY
(Nr. Daventry)
HASELEY
Nr Warwick
7
HEMPSTEAD (1573)
(nr. Saffron Walden)
HEMEL-HEMPSTEAD
EASTHAM
(1580)
STONOR PARK
Oxon.
(1581)
WANDSWORTH
(1572)
1
LONDON
(1572)
2 EAST MOSELEY
(1588)
© W.K. SESSIONS

III

SECRET PROVINCIAL PRINTING

As stated on page 1 of the Introduction, Queen Mary's Charter of 1557 to the London Stationers Company confined English printing to members of that company except that Oxford and Cambridge claimed earlier rights, as will be described in later chapters.

Elizabeth I from 1558, saw fit to retain these restrictions, though in support of the Protestant rather than the Catholic religion, and this she did in 1559, the year after she became Queen. The restrictions were further strengthened by a Star Chamber decree in 1566 and they continued in the reigns of James I and Charles I. Thus supporters of the Roman Catholic faith (as also the more extreme Puritans who were entering the lists against the Bishops) were compelled either to print their controversial productions overseas, or in the separate Kingdom of Scotland, or to make use of secret presses in England. Some of these were located in London, as for example the undercover Quaker press at the Bull and Mouth Inn, turned Meeting House in the 17th Century (see W G Bittle's James Naylor: Sessions of York 1986). Others however, no doubt for greater privacy and therefore hoped-for safety, were situated either on the fringes of London as it then existed, or fled deep into the English or Welsh countryside.

William H Allnutt (the then Assistant Librarian in The Bodleian Library Oxford), wrote in the journal Bibliographica from 1900 a valuable series of articles on "English Provincial Presses". In great appreciation to him I now quote his deeply researched summary of this believedly first of the clandestine English provincial presses (Pages 157-161):-

The earliest of these secret provincial presses is connected with the name of Thomas Cartwright, and has been assumed to have been located at Wandsworth.

Cartwright, on account of his religious teachings, was deprived of the Lady Margaret Professorship at Cambridge in 1570, and of his Fellowship at Trinity College in 1571, after which he left the country for a while, going eventually to Geneva.

About this time the Bishops and the prevailing mode of Church-government were attacked in a little tract, which gained considerable notoriety, entitled, *An Admonition to the Parliament*. The authors of this tract being discovered to be two London ministers, John Feilde and Thomas Wilcox, they were confined in Newgate, and their imprisonment caused the setting up of the secret press with which, upon his return from abroad, Thomas Cartwright was connected, and from which issued the following:—

1 (*a*). [An introductory address, headed] Grace and peace from God. &c. [dated at end] From my chamber in London / this. 30. of September / in Anno. 1572. [2 leaves.]

(*b*). An Exhortation to the Byshops to deal Brotherly with theyr Brethren. [4 leaves.]

(*c*). An Exhortation to the bishops and their clergie to aunswer a little booke that came forthe the last Parliament / and to other Brethren to iudge of it by Gods worde / vntill they see it aunswered / and not to be caryed away with any respect of man. [6 leaves.] Sm. 8°.

2. A Second Admonition to the Parliament. [With an address] To the godly readers Grace and peace from God. &c. [36 leaves.] Sm. 8°.

Copies in the British Museum, Bodleian, and Trinity College, Cambridge.

157

3. Certaine Articles, collected and taken (as it is thought) by the Byshops out of a little boke entituled an Admonition to the Parliament / wyth an Answere to the same. Containing a confirmation of the sayde Booke in shorte notes.

The Prynter to the Reader.

Thys worke is fynished thankes be to God /
And he only will keepe vs from the searchers rod.
And though master Day and Toy watch & warde /
We hope the liuing God is our sauegarde.
Let them seeke / loke / and doe what they can /
It is but inventions / and pollicies of man.
But you wil maruel where it was fynished /
And you shal knowe (perchance) when domes day is ended.
Imprinted we know where / and whan /
Judge you the place and you can. J. T. J. S.

[8 leaves.] sm. 8°.

Copies in the Bodleian and Trinity College, Cambridge.

Soon after these tracts were issued there appeared, written by Dr. Whitgift afterwards Archbishop of Canterbury, *An answere to a certen Libel, intituled, An Admonition to the Parliament*, 1572, at the end of which the above productions of our secret press are also briefly noticed and replied to. A royal proclamation was also issued, dated 11th June 1573, by which the *Admonition* and all other books made for its defence were declared seditious, and their circulation prohibited. In spite, however, of Masters Day and Toy, and the hue and cry after his press, Cartwright succeeded in producing two editions of the following book, the second of which appeared after the abovementioned proclamation.

4. A Replye to an answere made of M. Doctor Whitegifte agaynste the Admonition to the Parliament. By T. C. 4°.

Copies of both editions are in the British Museum and Bodleian.

In his address, contained in the first edition, the Printer laments 'that the noble Science of Printing . . . should be readyer to wayte vpon the defence of corruptions, then vpon the sinceritye of the trueth,' and regrets that for want of wealth 'to furnishe the Print wyth sufficient varietie of letters' he is compelled 'to vse one letter for three or foure tongues.' He also speaks of his 'want of long training vp in thys mysterie' and of his not being near the author, who only came to him 'when the halfe of the booke was Printed.'

In the second edition the Printer's address is entirely new. He says: 'Some perhaps will maruel at the newe impression of thys boke . . . notwythstanding our most gracious Princes late published proclamation / procured rather by the Byshops / then willingly sought for by her maiestie,' . . . and gives his initials at the end as J. S.

These initials are doubtless correct and represent John Stroud, formerly a Minister at Yalding in Kent, who is known to have had a hand in the production of Cartwright's *Replye*, for which he was called up and examined before the High Commission. But the names of two other printers are also connected with this Press. Bishop Sandys, writing to Lord Burghley, says: 'I have caused to be found and taken in the country a printing-press, with the whole furniture; the printer, called Lacy, with certain others of that confederacy, are also apprehended. They have printed Cartwright's book again, in a fair print, to the number of one thousand, as Lacy voluntarily confesses.' Archbishop Parker, also writing to Lord Burghley in November 1573, says: 'the warden of the printers, Harrison . . . told me that one Asplyn, a printer to Cartwrightes boke, was after examination suffred agayn to go abrode, and taken in to service in to Master Dayes house, and purposed to kyl hym, and his wyf, &c., and beyng asked what he ment, he answered "The Spryte moued hym," so . . . taken and in preson. . . .'

In the *Stationers' Registers* occur two printers named Asplyn; Robert, admitted a freeman, Oct. 5th, 1569, and Thomas, bound to John Day for eight years on March 25th, 1567. The latter is probably our man, a runaway apprentice, discovered and taken back by his master.

And now as to the locality and capture of the press. It has generally been supposed, as before stated, to have been worked at Wandsworth, and this may be partially correct; but there is no doubt that, if so, it was removed thence, and is identical with one seized on the 26th of August 1573, as recorded by Professor Arber in his *Transcripts of the Stationers' Registers*, I. 467-9.

Item laide owte the xxvjth of Auguste 1573 for my [John Harrison's] Iorney to Hempsteade with the pursevaunt.	xix^{s}.
Item to the Carter yat brought the presse	x^{s}.
Item to ye Constable of Hempsteade for bringing vp ye men	viij^{s}. vj^{d}.
Item for nailes to naile vpp the chestes	xij^{d}.
Item to the pursevaunt for his paines	xxx^{s}.
Item receaued of Henrie Bynneman for wearing the letter [*i.e.* for the use of the type] that came from Hempsted	xv^{s}.

Professor Arber has taken Hampstead to be the place where the press was seized; but from the costs of the journey and the amount paid to the pursuivant, compared with other similar entries in the accounts of the Company of Stationers, it would seem that some place more distant is intended.

This may be either Hemel-Hempstead, Herts, or Hempsted, near Saffron Walden, Essex.

The amount of fifteen shillings paid by Bynneman for 'wearing the letter that came from Hempsted' is repeated in the next year's accounts, and the use to which the type was immediately put is most ironically curious.

In 1574 appeared a fairly thick volume of small folio size, *The defense of the Aunswere to the Admonition, against the Replie of T. C. By Iohn Whitgift, Doctor of Diuinitie. Printed at London by Henry Binneman, for Humfrey Toye.*

In this volume the whole of Cartwright's *Replye* is actually reproduced in the Hempsted fount, Whitgift interpolating his own *Defense*, after each successive section, in a larger and more imposing black-letter. Binneman, however, was able to print Cartwright's Greek quotations, etc., in their proper letter, and to dispense with slanting lines in favour of commas.

The controversy between Cartwright and Whitgift did not end yet, but we are not entitled to follow it further here. A warrant was issued Dec. 11, 1573, for Cartwright's arrest, but he escaped the country and did not return till 1585, after which he evidently moderated his religious views, and died unmolested, December 27; 1603.

In 1602 Robert Waldegrave, by then King's Printer to James VI in Scotland, printed Thomas Cartwright's "long delayed book": <u>With God in Christ</u>; <u>The Answere to the Preface of the Rhemish Testament</u>. (See the end of Chapter IV on Robert Waldegrave following).

B THE CATHOLIC RECUSANT PRESS OF FATHERS ROBERT PARSONS and EDMUND CAMPION: 1580-1581

In 1580 the Society of Jesus asked Robert Parsons (or Persons) and Edmund Campion (Campian) to lead a small disguised group back from the continent to Elizabethan Protestant England on a dangerous mission to make contact with many who continued in private to uphold the old Catholic rights of worship and belief. Although Father Campion was questioned at Dover, the group reached London safely.

They soon aquired a small press, types and materials with Stephen Brinckley in charge, described as a Batchelor of Civil Law, and this was secretly set-up in a house called Greenstreet belonging to Edward Brookesby, (or Brookes) whose son William was active for the Catholic cause. Allnutt comments:-

> "This house has been variously placed at five, six and seven miles out of London, and is doubtless identical with Greenstreet House, East Ham, Essex, now [1900] known as Boleyn Castle.
>
> The following three books, written by F. Parsons, were probably printed here :—
>
> 1. A brief discours contayning certayne reasons why Catholiques refuse to goe to Church. Written by a learned and vertuous man, to a frend of his in England. And dedicated by I. H. to the Queenes most excellent Maiestie. [Monogram I. H. S. with a cross above and three nails below, all within a border of flames.] *Imprinted at Doway by John Lyon*, 1580. *With priuelege.* Sm. 8°. [The Dedication is written in the name of John Howlet.]
>
> 2. A Discouerie of I. Nicols minister, misreported a Iesuite, latelye recanted in the Tower of London. Wherin besides the declaration of the man, is contayned a ful answere to his recantation, with a confutation of his slaunders, and proofe of the contraries, in the Pope, Cardinals, Clergie, Studentes, and priuate men of Rome. There is also added a reproofe of an oration and sermon, falsely pretended by the sayd Nicols to be made in Rome, and presented to the Pope in his Consistorye. Wherto is annexed a late information from Rome touching the autētical copie of Nicols recantation. [Monogram, with quotations on each side and beneath.] Sm. 8°. [Without imprint.]
>
> 3. A brief censvre vppon two bookes written [by William Charke and Meredith Hanmer] in answere to M. Edmonde Campions offer of disputation. [Quotation from] Deuter. capit. v. ver. 5. [Monogram.] *Imprinted at Doway by Iohn Lyon*, 1581. *With privilege.* Sm. 8°.
>
> Exceptionally fine copies of the above three books are in the Bodleian Library.

From subsequent questionings, torture and trials it is known that there were initially a practical team of seven, and that they were all 'dressed as gentlemen' to seek to avoid discovery of the press' whereabouts. However one of them (unstated), while in London procuring more paper, was seized, imprisoned and put to the rack; and although he is said to have divulged nothing, the printing group scattered and also removed the press, secretly, further out from London.

"It has been stated that Jesuit Father Parsons' work Brief Censure* was printed at Henley Park in the house of Francis Browne, brother of Viscount Montague [and] Parsons himself, who had to be constantly on the move [and in disguise] was certainly being entertained there during the writing of it".

Thus the press moved secretly and successfully either via Henley Park or direct to its final location a few miles north of Henley, namely Stonor Park, where the Stonor family offered part of their house for this secret recusant press. Stonor Park is a large house in Oxfordshire some forty miles from London, which is still occupied by Thomas Stonor, the 7th Lord Camoys and his family. It was and is beautifully and discreetly situated in some 240 acres of rolling parkland, with herd of fallow deer on the lovely wooded yet open slopes. Stonor Park is about five miles north of the River Thames, leading two of the Campion biographers to speculate that it would be safer and easier for the press and all its varied necessary equipment and supplies to be brought up river rather than by road.

* See Facsimile

The Administrator of Stonor Park, Derek Boddy, kindly informs me that in this 1580 period, "the estate was owned by Francis Stonor (later knighted). He lived at Blounts Court about five miles from Stonor, leaving the occupancy of the house to his younger brother, John. It is recorded that their mother, Cecily, paid substantial recusancy fines and whilst she spent a great deal of time at Stonor, she lived in a lodge house. There can be little doubt that she was instrumental in harbouring priests at Stonor, but the organisation was by John Stonor who made the arrangements in London for the transfer of the printing press to Stonor, and for the provision of the labour to assist Brinkley with the printing." Afterwards both Cecily and John suffered house arrest for complicity and John was committed to The Tower of London, but later released.

In footnote 1 of Allnutt's following quotation (overleaf) he confuses two Francis Stonors, father and son. Cecily was the wife of Sir Francis Stonor (1520-1564), and it was their eldest son, Sir Francis, who was "High Sheriff of Oxon", and who lived from 1553 to 1625.

The fourth (possibly fifth) and final extant production of this secret press Rationes Decem* or "Ten Reasons for renouncing the Protestant and embracing the Catholic Faith", was thus printed at Stonor Park with open authorship 'Edmundus Campianus' and with the magnificent colophon 'Cosmopoli 1581'. Allnutt records its title as follows

4. Rationes Decem: Quibus fretus Certamen adversariis obtulit in causa Fidei, Edmundus Campianus E Societate Nominis Iesv Presbyter. Allegatæ ad clarissimos viros, nostrates Academicos. [Monogram with quotations from Luke 21, and Ps. 63.] [Colophon:—] Cosmopoli 1581. Sm. 8°.

* See Facsimile

This is STC[2] 4536.5: the entry refers to it as "8° in 4s. [Stonor Park, Henley-on-Thames, Greenstreet House Press, 1581]". Extant copies are recorded at five locations namely: Campion Hall, Oxford; St Edmund's College, Ware, Herts; Marquess of Bute; Bamburgh Castle on deposit at University of Durham; and Stoneyhurst College, Whalley, Lancs. I was privileged to visit Catholic Stoneyhurst College in 1977. It is a storehouse of early Catholic books, often written and produced in conditions of difficulty and danger, and so deliberately lacking in specific imprint information.

Allnutt quotes from some autobiographical notes, extant at Stoneyhurst, written in 1598 by Father Persons [Parsons] (who survived) as follows:-

'Having received F. Campions book de 10 rationibus compounded by him in Lancashire I sent for him to come up and print it himself: and in the meane space I procured another place for a print, to witt a house of my lady Stonar[1] standing in a wood fast by Henly, where Mr. Brinkly also was prefect of the printers and Mr. Wm. Maurice priest was procurator to buy the paper and other necessaries: and F. Campion coming up went to lye there and so printed his book. And after his book was printed & the most parte of the copies sent to Oxford against the Act,[2] which was in the beginning of July, F. Campion and I departed from that house about the 10th August[3] being Munday and upon Sunday after he was taken at Lyford by treason of Eliot, as in his life is sett down.

'I lying in Henly-park at Mr. Fran. Brown's house heard of the taking and passing by & sent Rob. Alfield my man to see him, and within a month after was Mr. Brinkley taken with all the printers and print and carryed bound to London.'

[1] Cecilia, wife of Sir Francis Stonor, High Sheriff of Oxon, 159⅗.

[2] They were secretly distributed to the extent of over four hundred copies in St. Mary's Church, where the Act (or as we now call it the Encaenia or Commemoration) was then held.

[3] Parsons, writing these notes from memory eighteen years later, mistakes the month, which must have been July.

Evelyn Waugh, in his Edmund Campion biography writes (p 146)

> At first sight the little volume—it is barely 20,000 words in length—shows little evidence of the difficulties under which it was produced. It has an elegantly spaced title page, decorated with a sacred emblem; the press-work is regular and the composition free from misprints. An expert examination has revealed certain peculiarities. Since the work is in Latin, Roman type had been used (Persons' English tracts were in black letter), but the printers were working at the disadvantage of great poverty of materials; after the first pages the dipthong "Æ," which occurs frequently, runs out, and is replaced by the italic "*Æ*," by "Œ," and even by "E." These substitutions become more frequent as we approach the end of signatures C, H and I, while at the beginning of the next signature the fount Æ reappears, suggesting that the sheets were printed off and the type distributed and reset three times during the printing. There is no Roman query sign; black letter is used in its place. There is no Greek fount; Campion's Greek quotations have to be given in Roman italics. The watermark is the common English pot. It occurs irregularly in the first, fifth and tenth gathering. The sheets run from A to K in fours, 16-mo.; there are forty-four folios; that is to say, that only eight pages went into each printing frame; this gives us a frame that could be covered by a half folio sheet, nine inches by thirteen.

On Easter Sunday 1988, with my wife and daughter's family, I had the very great pleasure of visiting this lovely secluded Stonor Park where visitors received a personal welcome from the seventh Lord Camoys (born 1940).

From the first floor a steepish flight of about 20 steps goes up to a Saint Edmund Campion room with historical displays in an area now known as Mount Pleasant. From there one can look through a recently-made window into the reconstructed room where the secret printing was produced, a room of perhaps 20 x 15 feet amongst ancient black timber rafters.

However in Campion's time neither of these upper floors existed over the main hall. In 1581 the location of the secret press was much more spectacular, much more secret.

On the ground floor a large fireplace had two separate chimney flues. In fact one of these took all the smoke, and the other flue was relatively free of smoke and soot.

So, like young Tom in Charles Kingsley's Water Babies Edmund Campion and the practical printing recusant team needed to scramble grubbily up the cleaner chinney flue to reach the totally windowless space amidst the attic rafters. This likewise was the means of ingress for the paper, ink, type and above all for the printing press; indeed this short-lived Stonor print-shop must claim a global all-time record for narrowness and vertical difficulty of the entry and exiting arrangements.

Clearly, only a very small printing press, in separate parts for attic-construction, could negotiate even a medieval chimney flue. Being largely of wood, it was fortunately relatively light but the stygian climbing difficulties remained.

The actual chimney no longer exists, but the already quoted extract from Evelyn Waugh's Campion biography states the relatively small sheet size needed to produce Decon Rationes namely "nine inches by thirteen", and it is conceivable that sections of a small wooden hand press just big enough to print such a sheet-size could have been man-handled (priest-handled) up a medieval-size chimney flue. Perhaps some other printing student (even more diligent than me!) may in the future like to assemble the main parts of such a small reconstructed press at the foot of a large chimney flue in some other stately home, with the object of making a trial of the practical difficulties encountered and surmounted by this resourceful secret printing team.

Campion himself had left Stonor before being betrayed and captured; but shortly afterwards protestant poursuivants descended upon Stonor and took into custody the rest of the printing team. However, as Derek Boddy, Stonor Administrator, kindly told me, even then the actual miniature printing press was not discovered; and remained hidden in the flue-access attics for many years afterwards, though sadly the apparatus does not exist today.

Although the 1581 "Ten Reasons" book "took several weeks to print off...., it was ready in time for Commencement at Oxford, Tuesday June 27th."

Louise Imogen Guiney in her Blessed Edmund Campion gives the most lively description of this occasion, thus:-

> The first copies bound, about four hundred in number, were hurriedly stabbed instead of stitched... When the [Oxford undergraduates] entered the Church of St Mary the Virgin, the benches were found littered with the 'seditious' books. Their dedication was 'to the studious Collegians flourishing at Oxford and Cambridge', and the youths in question were just in the humour to read them; and read them they did, then and there, instead of attending to the important annual function.......

Subsequent to that famous or outrageous occasion in Oxford four hundred and seven years ago, there have been many ingeniously-planned "book publication launches". However, I ask any reader to be good enough to send me details of any earlier, equivalently dramatic 'book-launch-spectacular'* than this in Oxford in 1581. Needless to say this was an occasion when the author did NOT sit at a table autographing copies! To the establishment it was indeed a seditious book, resulting in only five copies being said to be extant.

In 1581 Edmund Campion was still well remembered and admired in Oxford for his 'golden youth' and early manhood as an orator, debater and thinker. He had become a B.A. twenty years earlier in 1561. When Queen Elizabeth had visited Oxford in 1566 Campion had been selected to welcome her in the name of the University, and earlier still while at St Paul's School, Campion had been chosen to speak words of welcome to Queen Mary "on her solemn entry into London". Useful details of Campion's gifted career are given in the Dictionary of National Biography in eight columns from page 850. This entry however gives the Decem Rationes imprint wrongly as being "Douay" and includes no details of the location of the secret press.

Stephen Brinkley and his four assistants (seemingly reduced from the original seven) namely John Harris, John Harvey, John Tucker, and John Compton were soon afterwards seized at Stonor Park. The last named recanted at the point of the sword and was liberated. The others were imprisoned, Brinkley for two years at which time he was discharged on bail but went abroad (travelling with Father Persons) and eventually succeeded George Flinton as Catholic printer at Rouen.

* (The "wooden spoon" for a "leaflet launch" that was planned but failed, was on the eve of the civil war battle of Edgehill near Banbury in October 1642 - See p 17 of my King's Printer at York and Shrewsbury.)

As on other occasions, it will be noted that the undercover practical printers were dealt with less harshly than the authors, for Edmund Campion (though he had left Stonor Park in time) was not treated so lightly. George Eliot "a renegade priest-catcher" informed on Campion. He was arrested in the house of Mrs Yates at Lyford, Berks., on 16th July 1581, confined and cruelly tortured on the rack in the Tower of London and then on 1st December of the same year he was hanged, drawn and quartered. He had repeatedly stated his loyalty to Queen Elizabeth but would not compromise in his open adherence to what was to him the true and original Christian religion. Although author rather than practical printer, he was part of the short-lived undercover printing team and so seems qualified to be counted as one of the patron saints of printing, for he was subsequently canonised by the Pope for his bravery in the Catholic cause.

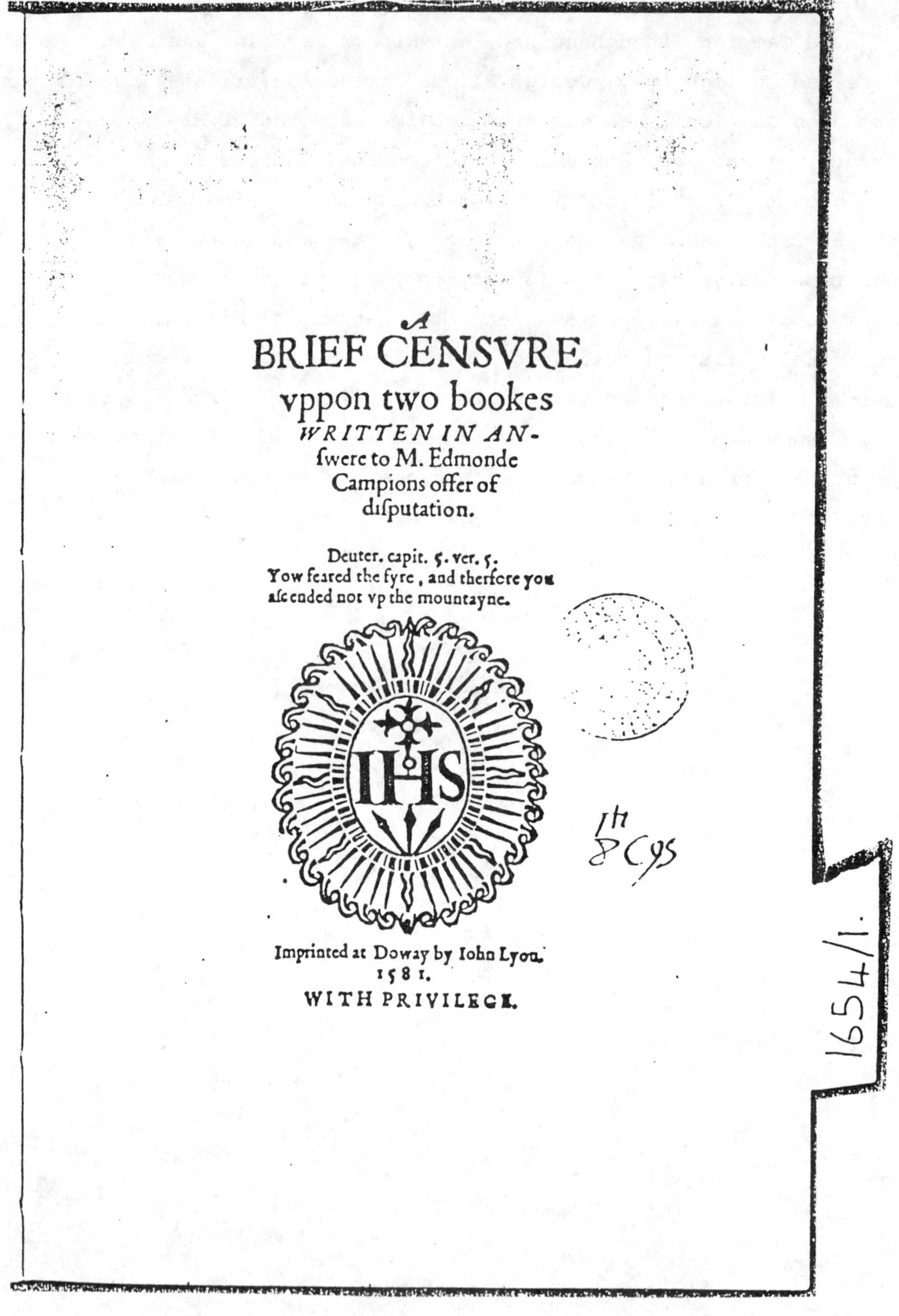
A
BRIEF CENSVRE
vppon two bookes
WRITTEN IN AN-
swere to M. Edmonde
Campions offer of
disputation.

Deuter. capit. 5. ver. 5.
Yow feared the fyre, and therfore you
ascended not vp the mountayne.

Imprinted at Doway by Iohn Lyon.
1581.
WITH PRIVILEGE.

Courtesy British Library London

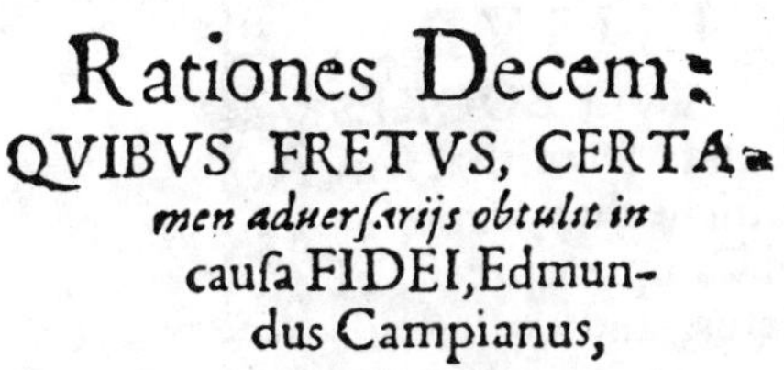

Rationes Decem:
QVIBVS FRETVS, CERTAmen aduersarijs obtulit in
causa FIDEI, Edmundus Campianus,

E Societate Nominis IESV Presbyter:
ALLEGATÆ
Ad clarissimos viros, nostrates Academicos.

Ego dabo vobis os & sapientiam

Psal.63.
paruulorum facti sunt plagæ eorum.

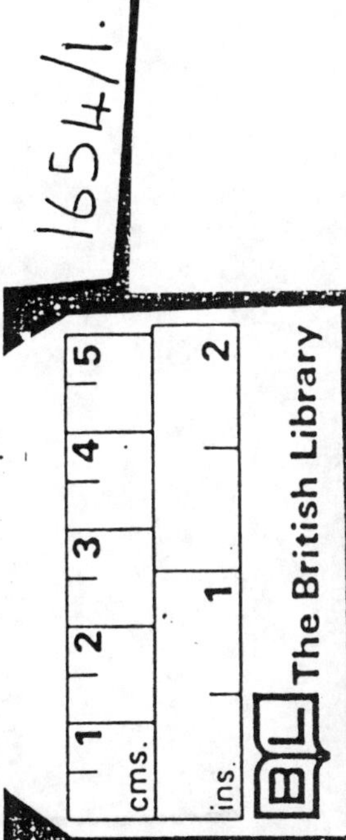

Courtesy British Library London

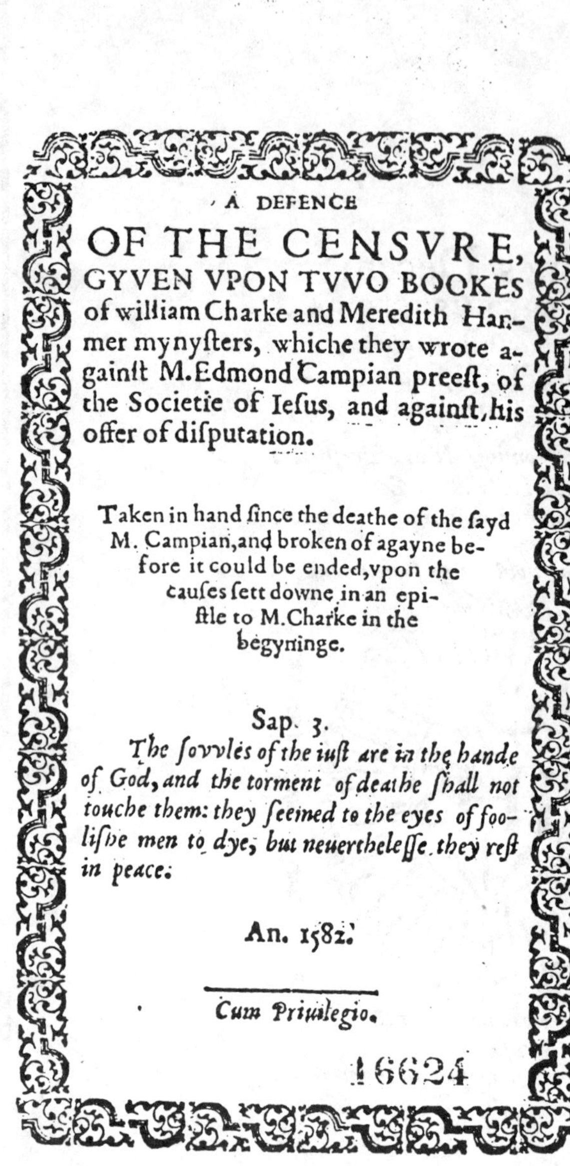

A DEFENCE
OF THE CENSVRE,
GYVEN VPON TVVO BOOKES
of william Charke and Meredith Harmer mynysters, whiche they wrote against M. Edmond Campian preest, of the Societie of Iesus, and against his offer of disputation.

Taken in hand since the deathe of the sayd M. Campian, and broken of agayne before it could be ended, vpon the causes sett downe in an epistle to M. Charke in the begyninge.

Sap. 3.
The sovvles of the iust are in the hande of God, and the torment of deathe shall not touche them: they seemed to the eyes of foolishe men to dye, but neuerthelesse they rest in peace.

An. 1582.

Cum Priuilegio.

Courtesy British Library London

Ierom.ep. 102.& in fine catalo. & corrected by S.Ierom? which was also so hyghlye cōmended by S.Augustin?what other better translation (I saye)hath william Charke than this auncient, which he so contemneth? except he will name some latter of our tyme, as of Erasmus, Luther, or the like: whiche Beza hym selfe notwithstandinge affirmeth to be nothing lyke the olde trāslatiō for exactnes. The fowerth poynt which he addeth is a shameles lye, that the septuagint in greeke doe dissent from the woorde *immaculata* in the latin. For their woorde is AMOMOS, which their owne lexicon will expound vnto them to be *immaculate, innocent, irreprehensible.*

Augu. ep. 10.ad hieron.

Præfat. in Nouum tes. an.1556.

AMOMOS.

TAMAM or TAM. To returne therfore in a woorde or two to the originall text, the hebrew woorde is TAMAM, or TAM which the septuagint doe interpret (as you haue heard) AMOMOS: that is, irreprehensible: and the auncient latin translation *immaculata*, immaculate. And what refuge then can M.Charke fynde heere? I doe not denye, but that it signifieth also, *perfect*: for that, what soeuer is irreprehensible and without spott, may also be called perfect as hath bene shewed. But how doeth this proue that it signifieth to be perfect in sense, in suche sorte, as it may not be wrested, or peruerted? In the 118. Psalme where our auncient translation hath *beati immaculati in via*: your owne englysh bible hath translated it (M. Charke) *blessed are those that be vndefyled in the vvaye:* and the Hebrew and greeke woordes are TAM, & AMOMOS, as in the other text. How then doe you rayle at our olde auncient translation for that wherein your new englishe byble doth the verye same? the lyke you may see in infinite other places: as leuit. 3, v.1. & 6. Also Num. 6. v. 14. VVhere sacrifices are appointed to be *immaculate*, according to the auncient translation. And your englishe byble translateth it so too, sayeinge they must be without blemishe: where the hebrew and greeke woordes are TAM and AMOMOS, as before.

Abourt the trāslation of immaculata.

An.1549

By whiche is seene, that M. Charke careth not whether he runneth, what he forgeth, or whome he reprehendeth, so he maye seeme allwa yes to saye somewhat:

what: And of all other shyftes, this is the last, and the easiest, and of most credit, and least able to be spyed of his reader, (as he thinketh) to inueighe against the olde latin translation, when he is pressed vnauoydablye with any place of scripture alleaged. For this shyft, besides the present couering of the difficultie, yeeldeth also some opinion of Learning to his Maister, gyuinge men to vnderstand, that he is skillfull in the learned tongues: whereas God knoweth, the refuge is vsed for bare miserie: as it alwayes appeareth, when it cōmethe to examination.

And this shall suffice for this sixt article.

HEERE the Authour vvas interrupted by a VVritte *de remouendo*, so as he could not for this present passe on any further: as more at large is shewed at the beginning, in an epistle to M. Charke.

Born in January 1540
captured at Lyford Grange in July 1581
— Hanged at Tyburn for his Faith in 1581.

Edmund Campion

Photograph: Focus, with kind permission of Colonel Dingwall

Reproduced by courtesy from the back cover of <u>A Kind of Warfare</u> by Juliet Dymoke (1981). This is a well-written historical novel about Stonor Park and family at the time of this secret printing, and is available for purchase (1988) from the Administrator, Stonor Park, Henley-on-Thames, Oxfordshire.

C LITTLE ORME, near LLANDUDNO, CAERNARVONSHIRE (Catholic): 1580-1587(?)

The first printing on Welsh soil (or should it be sand?) was in a sea-cave at Rhiwledyn, better known as the Little Orme headland, near Llandudno, in Caernarvonshire, now called Gwynedd.

In 1558 when Protestant Elizabeth replaced Catholic Mary on the throne of England, several Catholic scholars fled from Wales to Italy and other continental centres such as Rheims. One of these, Gruffydd Robert produced in 1567 the first part of a Welsh grammar Dosbarth Byvr, which was printed in Italy. In his foreword he writes (in Welsh): "Do not marvel at seeing so many errors in printing, because the men who were doing the work were not acquainted with the language and not bothered overmuch with the task, or easy to persuade.... working things out as they felt like it."

Robert Pugh and his family of Penrhyn Creuddyn were Catholic supporters and admirers of the writings of Gruffydd Robert; and it is thought that their project for establishing a clandestine press in a sea-cave in the Little Orme headland began in about 1580. Type, wooden hand-press, paper, ink and printing apparatus would most likely be conveyed to the cave by boat at night, guided to the shore by seaward shining lanterns, in best smuggling tradition.

The only book emanating from this unusually-situated press was Y Drych Christianogawl, in 1585, a work of some 180 pages, with a Rouen imprint. Gruffydd Robert then of Milan wrote the foreword. The title page contains a fair range of roman types in a variety of sizes and includes a Biblical quotation in black-letter type.

> "The text contains no letter W, two lower-case v's being used by placing them together for the purpose. This strongly suggests that the type was of French rather than English origin, which is claimed in an apology to the reader by Rosier Smith (a Catholic agent based on Paris, who was a former pupil of Gruffydd Robert and who may well have provided continental liaison and motivation for this amphibian typographic project). The book contains oddities of Welsh spelling, due it is said to lack of printing characters. Apart from problems of overall shortage, the frequency of usage of individual letters of the alphabet differs in the Welsh language, as compared with French or English."

"The finding of the [Little Orme] press is chronicled in the State Papers of Elizabeth I, in a letter dated April 16th, 1587, written by Dr William Griffith, a magistrate for the Caernarvon area. It is stated that a local informer had reported the discovery of some twelve priests working at a printing press in a cave at Rhiwledyn to the local Squire, Sir Thomas Mostyn; there is no incrimination of the Pugh family. On receipt of the report, Sir Thomas Mostyn led a party of forty men to the mouth of the cave, but did not enter, fearing an ambush (by the twelve priests?).
A guard was mounted by the entrance of the cave for the night - but by the following morning the priests had vanished!

"The magistrate reports that on entering the cave it was found to be lined with wood and equipped with an altar for worship and provisions. There was no sign of the printing press.

"The craftsmen-priests had ample time during the night to dismantle their wooden press, but the magistrate and his party did find on the beach below the cave pieces of leaden type which the priests did cast into the sea on their flight to the boats. What seems more likely is that the printers in their haste spilled or "pied" type from their cases in loading in the dark - printing type was far too valuable to cast into the sea with intent"

A clear inference of this account is that the discovery of the illegal press was delayed so as to permit escape by sea, perhaps because of sympathy for Catholicism in the locality. Robert Pugh, master of the maritime press, was one of those who escaped in the boats, "eventually making his way to Lancashire, where he had friends and sympathisers." Two years later, in 1589, the Marprelate Press was finally captured near Manchester, but that was a Puritan press, at the opposite end of the Christian spectrum. Seventeen years later, in 1604, came the Catholic first-fruit of the Lancashire Birchley Hall Press. One should therefore be on the look-out for any possible Lancashire catholic printing connection from Little Orme to Birchley Hall despite the 17-years interval.

(The above account is based with grateful acknowledgement on a very interesting 55-page booklet by Stanley I Wicklen privately printed by Llandaff College of Technology in Cardiff in 1969. Its title Princes and Printers derives from Prince Charles having been invested Prince of Wales in that same year, 1969).

D PRINTING THE MARPRELATE TRACTS (Puritan): 1588-89: East Molesey, Surrey: Fawsley near Daventry: White Friars, Coventry: Wolston Priory: Warrington, Lancashire: Newton Lane, Manchester.

It was in April 1588 (with the Spanish Armada attempted naval invasion of England to take place in the same autumn) that copies were circulated of an anonymous Puritan tract, which had been secretly printed in London by Robert Waldegrave. On 16th of that same month, his London press was seized on behalf of the Stationers Company and on 13th May an order was made 'that the said books shall be burnte, and the said presse, letters and printing stuffe defaced and made unserviceable.'

Despite the London seizure, Waldegrave managed to save part of his printing apparatus, which was conveyed to the home of Mistress Crane in London, and afterwards removed to the same lady's country house in East Molesey in Surrey.

It may have been Waldegrave who printed here in August 1588 his friend John Penry's further work A Defence of That* Which Hath Bin Written. I have inspected this small volume at the British Library London (their Ref. 4103 aa 14). It is of compact page size c. 5-1/4" x 3-1/4", incomplete but with 64 main text folios. Additionally by October Waldegrave had printed two further anonymous tracts both in the same typefaces. One of these was Demonstration,* actually written by John Udall, just removed as vicar of Kingston. Udall was subsequently sentenced to death in February 1590/1591, but in fact died of natural causes towards the end of 1592.

Also printed and published in October 1588 at East Molesey was the Epistle, the first of the Marprelate tracts.

* See Facsimile pages

The ridiculing style of the fictitious Martin Marprelate may be shown from the following paragraph taken from this October tract (a quarto):

> "Compiled for the behoofe and overthrow of the Parsons/Fyckers/and Curats/ that have learnt their Catechismes/and are past grace: by the reverend and worthie Martin Marprelate gentleman ------"

The secret press moved next to the house of Sir Richard Knightly, at Fawsley near Daventry, where a further tract Epitome* was printed in November 1588. The next printing location was the house of John Hales called White Friars at Coventry, where Waldegrave printed three further pieces: Mineralls at the end of January 1588/1589, and A View (Supplication) by John Penry (see later) in February. Then towards the end of March 1589 the fourth Marprelate work was published: Hay Any Worke for Cooper.* This challengingly recorded the author and imprint: "Penned and compiled by Martin the Metropolitane. Printed in Europe/ not farre from some of the Bounsing Priestes."

By now the poursuivants were obviously 'closing in' on the authors, printers, binders and distributors. Consequently in early April 1589 Waldegrave, alarmed for his personal safety, fled the country, believedly taking with him the black-letter types used for printing the latest quarto tracts.

After Waldegrave's flight from White Friars at Coventry another printer John Hoskins (Hodgkins or Hodgkys) replaced him, assisted by Valentyne Symmes and Arthur Thomlyn. (These last two, as distinct from Hoskins, are traceable in the Stationers' Registers).

* See Facsimiles

The clandestine press now removed to the house of Roger Wigston, Wolston Priory, situated midway between Coventry and Rugby, where two further books were printed in small 8vo: the first Mar-Martine was issued in May 1589, with the second Wolston work, Theses Martinianae or Martin Junior, appearing in July "Printed by the assignes of Martin Iunior".*

At the very end of July this truly peripatetic press moved further north, to Warrington in Lancashire "probably Hodgkin's home", where, during unloading, type was spilled on the ground which probably resulted in their imminent capture. Nothing extant was printed at Warrington, and it was finally at a house in Newton Lane, Manchester, that the printers were surprised and captured by the Earl of Derby, on 14th August 1589. At the time of seizure they had printed about six quires of one side of a book (which of course never came out) entitled More Worke for Cooper.

The life of the press was indeed short, from Spring 1588 to August 1589, during which time it was in production in no less than six (and conceivably seven or more) locations. Yet even after the capture in or near Manchester, some of the type escaped seizure because about a month later there was gallantly printed, believedly in the house of Job Throckmorton at Haseley near Warwick, the last of the Martin Marprelate tracts, Protestatyon, a small octavo bearing "remarkable signs of the scarcity of types available and of the inexperience of the printer".

* See Facsimiles

E THE AUTHORSHIP OF THE MARPRELATE TRACTS

That these were printed in sundry specified places in England, outside London, is not open to doubt. As to their authorship, this was of intense and unrelenting enquiry at the time through poursuivants at the order of the ecclesiastical authorities and the Star Chamber and thence through the records of immediately following court cases. Since then and through to the 1980s the Marprelate authorship has continued through the four hundred years to be the subject of scholarly research, speculation and published writing to a degree unparalleled by any other series of books, save perhaps for writings on the authorship(s) of the books of the Bible or of the plays of Shakespeare.

It is (perhaps fortunately) outside the scope of this printing study, to attempt any new independent assessment of the hidden authorship of this Marprelate series of politico-religious satires of the late 1580s. In subject matter they were libellously anti-episcopal, and, beyond that, outspokenly critical of alleged shortcomings and misdeamenours both as to the lower rankings of the Church of England and with considerable venom as to Archbishop Whitgift of Canterbury - being attacks of a far more direct and virulent character even than the attack of the winter of 1987, with its tragic ending, into the anonymous preface of the latest Crockford annual.

The objective of these Mar-Prelate tracts was indeed to "mar the prelates" and the method chosen was a unique idiosyncratic mixture of direct libellous hurtful wording against some Church of England fundamental principal: (notably the episcopy system), and in great detail scurrilously against bishops and archbishop, by name and by nickname. Thus Archbishop Whitgift of Canterbury was derisively named as "Canterbury Pope" and "My Lord of Cant", and again "His Gracelessnes of Cant."; while Bishop Cooper of Winchester actually appeared by name in two of the Marprelate titles; namely Hay Any Worke for Cooper and the "work in progress when captured" title: More Worke for Cooper.

John Bridges, Dean of Salisbury, was also the subject of a virulent attack not only on himself personally but also on his wife's alleged dalliance with "Master Dr Day, sometime of Magdalins". He too appears in the top line of the _Epitome's_ title, _namely Oh read ouer D. John Bridges/ for it is worthy worke:_

These libellous named attacks, couched in vivid easy-to-read wording, with cleverly invented words, were interlaced with derisive anecdotage such as the Marprelate story of "Sir Jefferie Jones, an incompetent clergyman in Corley, Warwickshire, who imbibed too much ale. --- [He] made a rash vow never to go to the alehouse again [but] unable to forego his morning draught ---- he hired a man to carry him on his back to the alehouse" (_Martin Marprelate Gentlemen_ by L. Carlson, p 263).

Or again the Marprelate puncturing of the allegedly egotistical writings of Dr Richard Some, Master of Peterhouse, thus:

> If you give good eare, nothing els, I warrant you, but
> My sermons, My writings, My reasons, My arguments; and
> al is My, My, My ------ (Carlson p 290)

These secretly printed results, emanating from a series of secret press locations northward through England, were (and still are) a readable and heady mixture of scurrilous and jokey writing, which are possibly the first well-known examples of the writing of satire in the English language in that year of 1588, during the very summer and autumn of the fear-ful seabourne invasion by the Spanish Armada in July 1588, defeated in part by the experienced sea-going courtier Francis Drake and his fellow national-hero privateer colleagues, and in large part also by a succession of summer storms and gales up and around the shores of England and Scotland - those unpredictable bouts of rough weather which have protected these islands from actual or planned invasion on more than one occasion. (It cannot be often that a writer blesses our English weather with its difficult-to-forecast variability).

These undercover English-printed booklets, with their scurrilously readable mixture of humour and personal attack on respected Church of England dignitaries, were bravely distributed secretly into the large population centres, notably London, where they were passed surreptitiously from hand to hand, even into high places. There no doubt the nicknames and anecdotage were rapidly from mouth to mouth with that characteristic zest of not so homo sapiens for causing hurt to others. Indeed to the reading public of that autumn and winter of 1588 the avid under-cover reading of the Marprelate books plus the subsequent coffee house passing on of revealing or scurrilous tit-bits must have been akin to the equally avid under-the-table selling and reading of the then-banned-in-Britain Spycatcher M15 broken-promise revelations by author Peter Wright safe in Australia; whereas of those associated with Marprelate, Throkmorton and Strange were indicted in English courts while John Penry was to be hanged in 1593 for the writing of a subsequent book.

In that year of 1588 it was greatly to the honour of the religiously-sincere Puritan author John Penry that he stated he would have no part in personal or hurtfully joking attacks, because they were sinful, even though he, like Marprelate, was setting out puritan principles in his writings. This declared-at-the-time John Penry attitude towards the Marprelate mode of writing is one of the crucial indices in the search by others of the Marprelate authorship: to which subject I will shortly return.

The detail and substance of the Marprelate attacks on the high ranks of the Church of England were such that it was little wonder that reaction was swift and vigorous. Not only were under-cover investigators or poursuivants set to work with speedy and dramatic results, but also during that 1588/9 winter and spring there were a series of anti-Marprelate Plays (Present-day Archbishops and Synod Committees may care to ponder a revival of this public relations vehicle last used by the Billy Graham one-time ownership of one of the London Shaftesbury Avenue theatres). Furthermore Marprelate provoked a series of anti-Marprelate tract publications.

So what of the authorship of these uniquely idiosyncratic Marprelate anti-episcopal puritan satires?

It was in the summer of 1985 that my wife and I were researching for a few days in the botanically, bibliographically and ornithologically superb surroundings of the Henry E Huntington (railroad king) Library in San Marino, California, while resting at the comfortable Atheneum CalTech graduate facility within walking distance nearby. During this memorable visit I had the good fortune to meet Leland H Carlson, who kindly expressed interest in and made helpful suggestions about my 'green-back' series of then 6 and now 9 studies into the spread of British printing in the first two hundred years after Caxton. Somewhat naturally we touched on the subject of the printing locations of the series of Marprelate satires, and he directed me to the tabular summary of this which he had included in his 1981 publication, at that time undiscovered by me. I was indeed able to rectify that ommission then and there by purchase at the Huntington Library of Martin Marprelate, Gentleman: sub-titled Master Job Throkmorton laid open in his colours.

This is an extremely thoroughly researched work of 445 + 21 pages including appendices, detailed notes and extensive index which gives "chapter and verse" of the great differences between the writing styles, known-personalities and attitudes to specific topics as between (a) the Marprelate satires (b) the known and almost always openly printed John Penry writings and (c) the all-but-one undeclared writings of Job Throkmorton (1545 to 1601). This analysis is very explicit in quoted comparative examples, very readable and excellently printed in a commendably legible typesize - and personally I endorse the compelling build-up to his conclusion that Job Throkmorton was the author of the Marprelate series of satires rather than the previous alternative contender, John Penry or, as he titled himself in his own books, 'Penrì'.

Throkmorton was contemporaneously known as a learned, mature man with a facetious and biting tongue, with the status of a gentleman living comfortably at Haseley near Warwick. A one-time Member of Parliament, he had spoken in the House so bitingly about England's foreign policy that Lord Treasurer Burghley had ordered him to be forthwith imprisoned in The Tower of London; however, Throkmorton had characteristically got early news of Burghley's intention and had escaped in time to his Haseley estate. Throkmorton had hoped to use his biting power of words in the puritan cause in Parliament, but when that possibility failed he turned to the writing and undercover production of this series of venemous anti-episcopal satires.

By contrast John Penry was in 1588 but a young 25-year-old, an intensely serious, puritan evangelist who eschewed the devices of satire, hurtful humour and scurrility as being sinful and contrary to the new and true pure puritan religion. (see Carlson p 215).

It is surely ironic that whereas on the one hand Job Throkmorton, although indicted in Warwick in 1590, never came to trial and had his case left open (neither pro nor con) for all of eleven years until he died naturally and untried in 1601; yet Throkmorton had been indicted for "disgracing Her Majesties government" and specifically for the writing of scoffing and defamatory books under the pseudonym of Martin Marprelate (see Carlson p 268). In 1590 Throkmorton had made his submission to the Archbishop of Canterbury and to Lord Chancellor Hatton as well as to the Judges of Assize. My inference is that Throkmorton had indeed friends in high places.

As to the ardent young evangelist John Penry, although not the Marprelate author, yet he stayed in Throkmorton's house for 7 months from 2nd March to 2nd October in 1589. He probably helped with proof-reading and in other managerial ways. In stark contrast to the wily Throkmorton being let off the judicial hook, John Penry dropped in the noose at Southwark three years later in 1593.

However, Penry was not charged with the Marprelate writings; he was hanged for writing a 1590 book, published and printed by Robert Waldegrave in Edinburgh. This was judged to be a seditious book against the English magistrates. Its full title was: *A Treatise Wherein Is Manifestly Proved, that Reformation and those that Sincerely Favour the Same, Are Unjustly Charged to be Enemies, unto His Maiestie and the State.* Somewhat naturally subsequent scholars have shortened their references to this book so as to read *Reformation No Enemie*. Does not this long-winded and non-impact full title further endorse Carlson's and other's findings that the man (John Penry) who saw fit to "kill the human interest" in the important message of his final book with such a boring 'non-starter' title, was definitely *not* the same man who successfully titivated his prospective readers with such tempting titles as *Hay Any Worke for Cooper* with its superbly joking imprint "Printed in Europe not farre from some of the Bounsing Priestes."

Yet it was for writing *Reformation No Enemie* that the dedicated unworldly puritan evangelist John Penry was hanged in 1593 at the age of only 29. Surely it would have been tempting for the authorities to have at the same time "done justice to the other case" of but five years earlier; yet in fact at his 1593 death trial John Penry was *not* indicted for writing the Marprelate books. (Carlson p 312).

It should be added that in the *Dictionary of National Biography* the articles (both signed "S.L") on Penry and Throkmorton, clearly favour John Penry as the author of the Marprelate tracts, with John Throkmorton in the role of financier, organiser, and for a time 'Houser' of the clandestine press in his own home in Hasley, Warwickshire.

As to John Penry the *DNB* states that he "is reckoned by Welsh historians as the pioneer of Welsh noncomformity" and quotes contemporary records which describe him as "John of Wales".

Again in great appreciation of the excellent Martin Marprelate, Gentleman by Leland H Carlson (1981: Huntington Library, San Marino, California) I reproduce from its preliminary pages the following extremely useful excerpt.

A CHRONOLOGICAL OUTLINE

1545–1604

1545–1582

Throkmorton born in 1545. Oxford University 1562–66. "An Answer to Certen Peeces," *An Exhortation to the Byshops*, *An Exhortation to the Bishops*, *A Second Admonition to the Parliament*, *Certaine Articles*, all written by Throkmorton, 1572. "A Friendly Caveat to Bishop Sands," 1573

1583

6	July	Archbishop Edmund Grindal dies
23	September	John Whitgift confirmed as archbishop of Canterbury

1584

13	January	Throkmorton's letter to Ralph Warcuppe
23	March	Thomas Cooper confirmed as bishop of Winchester
	April	*A Dialogue concerning the Strife of Our Church*

1586

23	June	Star Chamber decree on printing X
4	November	Throkmorton's Speech on Mary Queen of Scots
		Throkmorton an M.P. in Elizabeth's sixth Parliament, 29 October 1586 to 23 March 1587

1587

23	February	Throkmorton's Speech on "The Low Countries"
27	February	Throkmorton's Speech on "The Bill and Book"
	February	Penry's first book published—*A Treatise Containing the Aequity of an Humble Supplication.* Presented to the House of Commons 28 February by Edward Dunn Lee
	March	Penry imprisoned in the Gatehouse for a month by the Court of High Commission
23	March	Elizabeth's sixth Parliament adjourned. Throkmorton absent throughout March, since he was in hiding
3	April	Throkmorton's petition to Lord Burghley
16	April	Easter
?	June	Dean John Bridges' *A Defence of the Government Established* published
?	November	Dudley Fenner's *A Defence of the Godlie Ministers* published. This is a reply to Dr. Bridges' *A Defence*

1588

	March	John Field, Puritan co-ordinator, dies. Buried 26 March
	March	Penry's *Exhortation* published, or possibly in February
7	April	Easter
16	April	Throkmorton's book, *The State of the Church of Englande* (also known as *Diotrephes*), confiscated in Waldegrave's shop X

xvii

16	April	Waldegrave's printing office raided
13	May	The Stationers' Company orders the destruction of Waldegrave's press, type, and equipment
	May	Penry's *Exhortation*, second edition, published
	May	Dr. Robert Some's first *A Godly Treatise* published
	June	John Udall removed as vicar of Kingston and silenced
	August	Penry's *A Defence of That Which Hath Bin Written* printed by Waldegrave
4	September	Robert Dudley, earl of Leicester, dies
5	September	Penry marries Eleanor Godley at Northampton
	September	Dr. Robert Some's second *A Godly Treatise* published
15–20	October	First Marprelate book, the *Epistle*, published at East Molesey
	October	John Udall's *Demonstration* published
10	November	Dr. Robert Some's sermon at Bartholomew Church in the Exchange. Throkmorton in attendance
14	November	The Queen commands Archbishop Whitgift and the Court of High Commission, with the aid of privy councilors, to apprehend the author, printers, and dispersers of Marprelate's *Epistle*
14	November	Depositions made by Cutbert Cook, John Good, Nicholas Kydwell, and William Staughton
15–20	November	Deposition of Stephen Chatfield
29	November	Examination of Walter Rogers
25–29	November	Marprelate's *Epitome* published at Fawsley
6	December	Giles Wigginton examined before the Court of High Commission

1589

1	January	Wednesday. Henry Barrow interrogated in the Fleet on his reply to Dr. Some
10 ca.	January	Thomas Cooper, bishop of Winchester, publishes his *An Admonition to the People of England.* First reply to Marprelate
26–28	January	Marprelate's *Mineralls* published at Coventry (not 20 February)
29	January	Penry's study in Henry Godley's home raided by the pursuivant, Richard Walton. Udall's *Demonstration* and Penry's manuscript reply to Dr. Some seized
4	February	Elizabeth's seventh Parliament assembles on Tuesday
8	February	Penry's *A Viewe (Supplication)* published at Coventry
9	February	Richard Bancroft's sermon at Paul's Cross. Denounces Martinists. Sermon expanded and published in March
13	February	Royal proclamation against Marprelate's books and broadside issued
15	February	Nicholas Tomkins examined by Richard Cosin, Master in Chancery
	February	Attempt to arrest Henry Sharpe at Northampton
2	March	Penry begins his residence as a guest of Throkmorton at Haseley, Warwickshire. Sought by pursuivants
20–27	March	*Hay Any Worke*, the fourth Marprelate work, published
30	March	Easter. Henry Sharpe consults with Sir Richard Knightley about confessing his own complicity in binding and selling Marprelate's books
4–5	April	Waldegrave departs from Coventry and Wolston
26	April	Andrew Perne dies at Lambeth
	May	Dr. Robert Some's third *A Godly Treatise* published
	May	Mar-Martine published

	June	Leonard Wright's *A Summons for Sleepers*, and *A Whip for an Ape: Or, Martin Displaied* published
13–14	July	John Hodgskin confers with Job Throkmorton and John Penry at Haseley Manor
14	July	Part of manuscript of *Martin Junior* found along the path
18	July	Job Throkmorton visits printers on Friday at home of Roger Wigston in Wolston
22	July	*Theses Martinianae*, or *Martin Junior*, printed. Printers leave Wolston for Warrington 29 July
29	July	*Martin Senior* published
	July	*Antimartinus*, a Latin treatise, published
1	August	Printers reach Warrington—probably Hodgskin's home
4	August	Cart and press arrive at Warrington. Type spilled on the ground. Probable cause for capture of printers
7	August	Printers move to Manchester. Press set up in a house
14	August	On Thursday, Hodgskin, Symmes, and Thomlin captured at Manchester while printing *More Worke for Cooper*
23	August	Saturday. Captive printers arrive in London
24	August	Hodgskin, Symmes, and Thomlin examined before the Privy Council
	August	"Martins Interim" completed. Not printed
	September	Henry Sharpe arrested—about 7–10 September
11–12	September	Hodgskin sent to the Tower for racking
10–15	September	Sharpe's preliminary examination
21	September	Summary Report on Marprelate and Penry and printers sent to Lord Burghley. Endorsed 21.7.1589 (Old Style)
20–30	September	Penry's *Appellation* published. Probably printed in July at La Rochelle. Delivered by Waldegrave to Throkmorton
20–30	September	Throkmorton's *Master Some Laid Open* published. Probably printed in July at La Rochelle. Delivered by Waldegrave
20–30	September	Throkmorton's *A Dialogue. Wherin Is Plainly Laide Open* published. Probably printed at La Rochelle in August. Delivered by Waldegrave to Throkmorton
20–30	September	Last Marprelate work, the *Protestatyon*, published
30 ca.	September	Second raid on Henry Godley's house in Northampton
2	October	Penry leaves Haseley and flees to Scotland
10	October	Symmes and Thomlin examined
15	October	Henry Sharpe examined by the commandment of Sir Christopher Hatton, lord chancellor
20	October	Sir Richard Knightley's first examination
20	October	*The Returne of the Renowned Cavaliero Pasquill of England* published. Possibly by Thomas Nashe
20	November	Sir Richard Knightley's second examination
25	November	John Hodgskin examined
27	November	John Hodgskin examined
29	November	Second examination of Nicholas Tomkins, by Dr. William Aubrey and Dr. William Lewin
10	December	Symmes and Thomlin examined before the lords commissioners
11	December	R. Jeffes' examination
11–12	December	Lawrence Jackson examined
	December	Hodgskin examined by the lords commissioners

1590

	January	Penry's *A Treatise (Reformation No Enemie)* printed at Edinburgh by Waldegrave

9	January	John Udall arrives in London, in response to a subpoena from the Privy Council. He left Newcastle-upon-Tyne on 29 December
13	January	John Udall examined at Lord Cobham's house in Blackfriars, by Lord Cobham; Lord Buckhurst; Lord Chief Justice Edmund Anderson of the Court of Common Pleas; John Young, bishop of Rochester; John Fortescue, chancellor of the Exchequer; Thomas Egerton, solicitor-general; Dr. William Aubrey; and Dr. William Lewin
19	January	Leonard Wright's *A Friendly Admonition* entered in the Stationers' Register
	February	The Puckering Brief compiled
13	February	Sir Richard Knightley, John Hales, Roger Wigston, and Mrs. Roger Wigston tried and sentenced in the Court of Star Chamber
	February	*An Almond for a Parrat*, probably by Thomas Nashe, was published in February or March
12	March	Richard Holmes examined
6	April	Sir Francis Walsingham, principal secretary, dies
ca.	June	Penry's *A Briefe Discovery* printed at Edinburgh by Waldegrave
3	June	Mr. Grimston examined
9	July	Humphrey Newman examined
24–25	July	John Udall arraigned before the judges of Assize at Croydon
1–5	October	Job Throkmorton indicted by a grand jury at Warwick
10–15	October	Throkmorton's letters of submission to Archbishop Whitgift and to the judges of Assize
14	October	Throkmorton's letter of submission to Sir Christopher Hatton, lord chancellor
6	November	Jenkin Jones, a kinsman of Penry, examined

1591

18–20	February	John Udall at the Assizes in Southwark. Sentenced to death on 20 February, but reprieved
	April	Throkmorton's court appearance—Easter and Trinity terms
	July	Penry in London because of the Hacket–Coppinger–Arthington scheme
28	July	William Hacket executed. Edmund Coppinger died the next day in prison
28	July	Penry flees from London to Scotland
18	August	Penry arrives in Scotland
26	August	Penry's third daughter, Safety, born
3	October	Henry Kildale, or Kyndall, examined. Waldegrave's assistant
20	November	Lord Chancellor Christopher Hatton dies
		Richard Cosin's *An Apologie: of, and for Sundrie Proceedings by Jurisdiction Ecclesiasticall*

1592

	June	Throkmorton's *A Petition Directed* printed by Schilders at Middelburg
	October	Penry arrives in London. Joins the Separatists
	December	Matthew Sutcliffe's *An Answere to a Certaine Libel Supplicatorie* published. Reply to *A Petition Directed*

1593

	March	Richard Hooker's *Of the Lawes of Ecclesiasticall Politie*
22	March	Penry captured in Ratcliffe, in Stepney

6	April	Henry Barrow and John Greenwood hanged at Tyburn. Friday
15	April	Easter
21	May	Penry arraigned before the Court of Queen's Bench. First indictment. Monday
25	May	Penry arraigned before the Court of Queen's Bench. Second indictment. Convicted and sentenced to death. Friday
29	May	John Penry hanged at Thomas à Watering, Southwark, Tuesday afternoon
		Bancroft's *A Survay*, his *Daungerous Positions*, and the Puritans' *A Parte of a Register* published

1594

	April	*The Defence of Job Throkmorton* published at Middelburg by Richard Schilders. Only book with Throkmorton's name
29	April	Thomas Cooper, bishop of Winchester, dies
3	June	John Aylmer, bishop of London, dies

1595

ca.	June	Matthew Sutcliffe's *An Answere unto a Certaine Calumnious Letter* published

1596

A Brief Apologie of Thomas Cartwright. The preface is by Throkmorton. Printed by Richard Schilders at Middelburg

1597

8	May	Richard Bancroft consecrated as bishop of London

1601

23	February	Job Throkmorton dies

1603

24	March	Queen Elizabeth dies. James VI of Scotland succeeds as James I. Waldegrave returns to England about April–May
27	December	Thomas Cartwright dies

1604

29	February	Archbishop Whitgift dies. Funeral solemnities 27 March
10	December	Richard Bancroft confirmed as archbishop of Canterbury

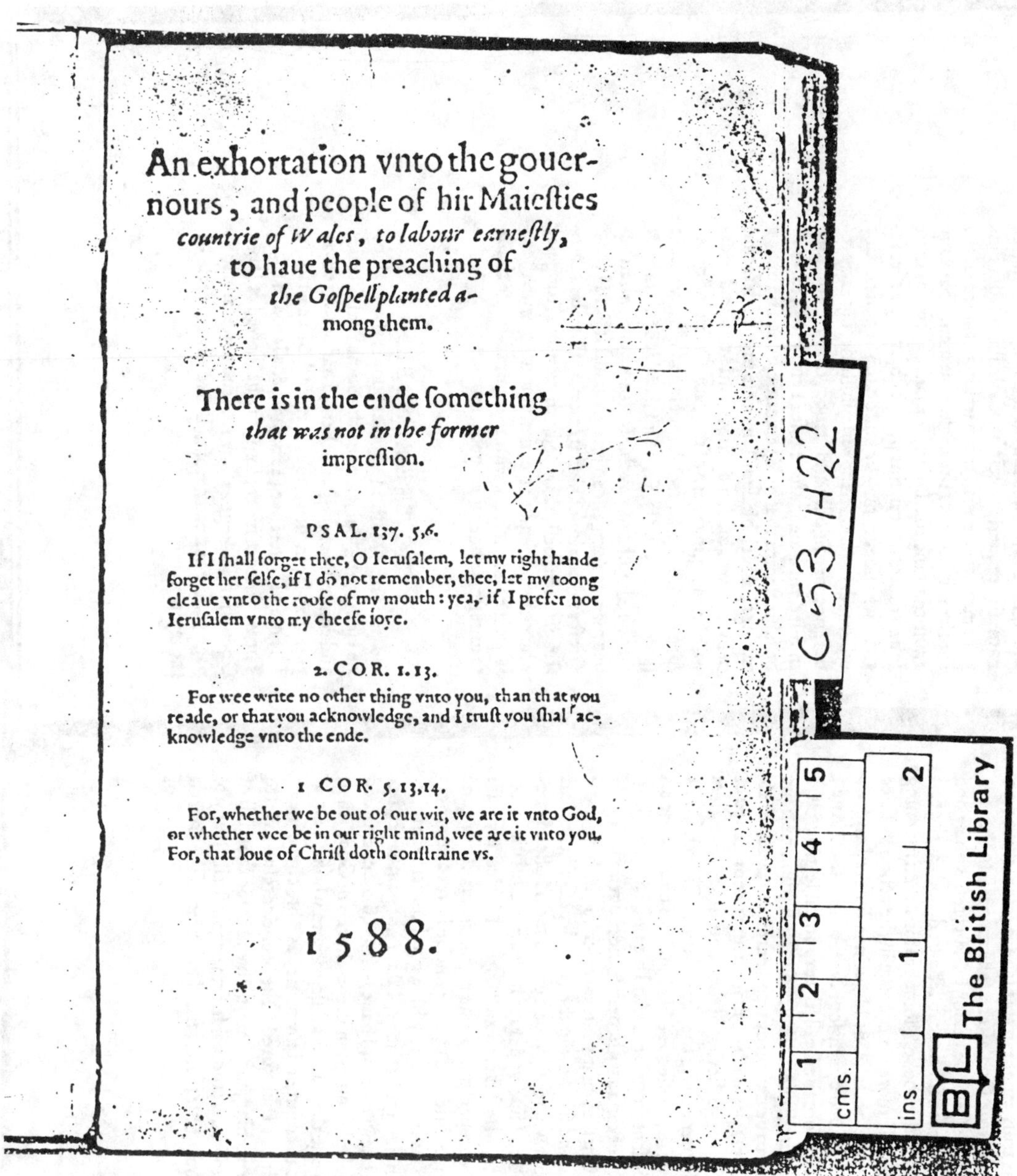

An exhortation vnto the gouernours, and people of hir Maiesties countrie of Wales, to labour earnestly, to haue the preaching of the Gospell planted among them.

There is in the ende something that was not in the former impression.

PSAL. 137. 5,6.

If I shall forget thee, O Ierusalem, let my right hande forget her selfe, if I do not remember, thee, let my toong cleaue vnto the roofe of my mouth: yea, if I prefer not Ierusalem vnto my cheefe ioye.

2. COR. 1. 13.

For wee write no other thing vnto you, than that you reade, or that you acknowledge, and I trust you shal acknowledge vnto the ende.

1 COR. 5. 13,14.

For, whether we be out of our wit, we are it vnto God, or whether wee be in our right mind, wee are it vnto you. For, that loue of Christ doth constraine vs.

1588.

<u>An Exhortation</u> by John Penry
See Carlson's listing: foot of page 84 (xvii)

Courtesy British Library London

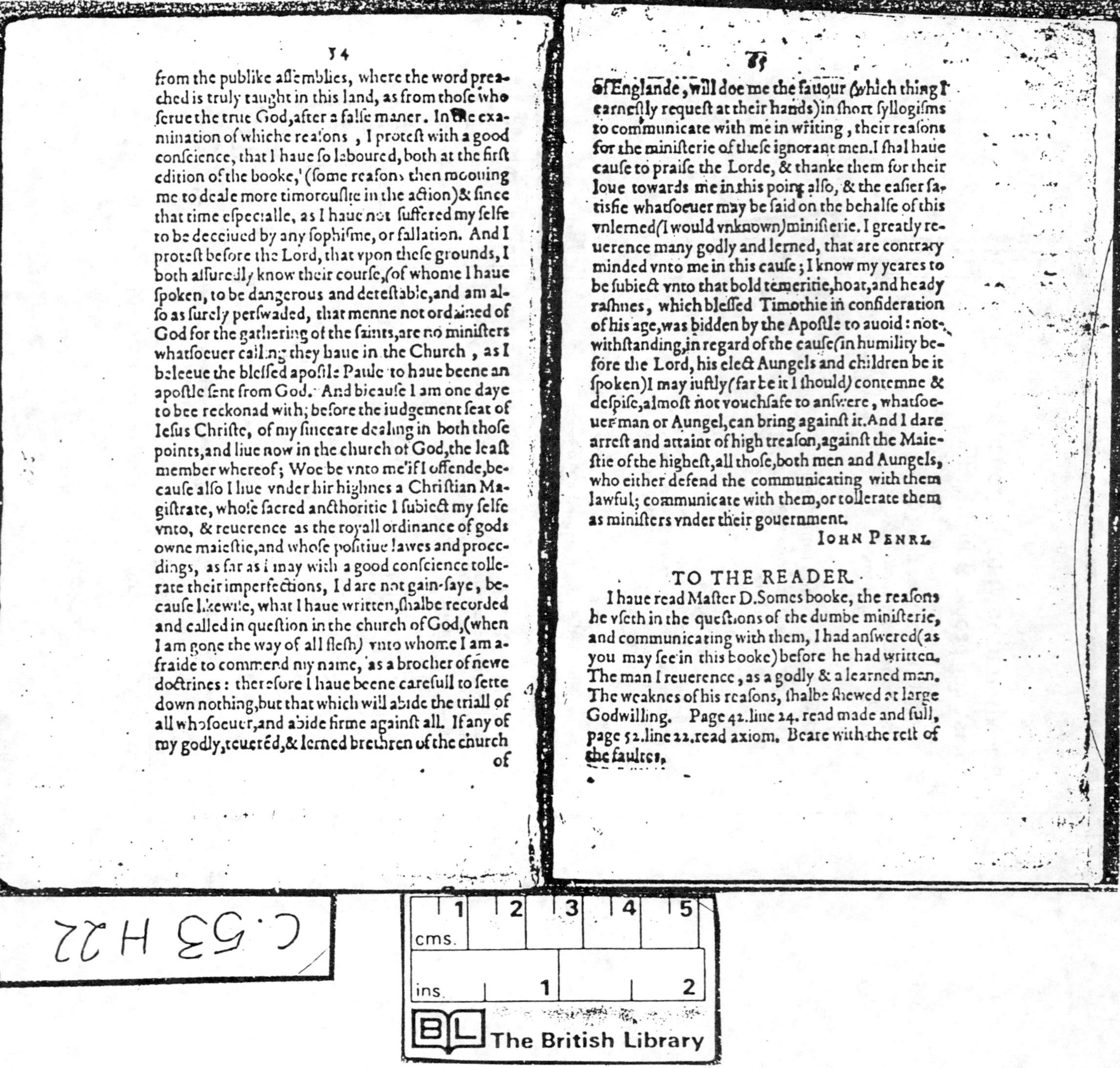

54

from the publike assemblies, where the word preached is truly taught in this land, as from those who serue the true God, after a false maner. In the examination of whiche reasons, I protest with a good conscience, that I haue so laboured, both at the first edition of the booke, (some reasons then mooving me to deale more timorouslie in the action) & since that time especialle, as I haue not suffered my selfe to be deceiued by any sophisme, or fallation. And I protest before the Lord, that vpon these grounds, I both assuredly know their course, (of whome I haue spoken, to be dangerous and detestable, and am also as surely perswaded, that menne not ordained of God for the gathering of the saints, are no ministers whatsoeuer calling they haue in the Church, as I beleeue the blessed apostle Paule to haue beene an apostle sent from God. And bicause I am one daye to bee reckonad with; before the iudgement seat of Iesus Christe, of my sinceare dealing in both those points, and liue now in the church of God, the least member whereof; Woe be vnto me if I offende, because also I liue vnder hir highnes a Christian Magistrate, whose sacred anthoritie I subiect my selfe vnto, & reuerence as the royall ordinance of gods owne maiestie, and whose positiue lawes and proceedings, as far as I may with a good conscience tollerate their imperfections, I dare not gain-saye, because likewise, what I haue written, shalbe recorded and called in question in the church of God, (when I am gone the way of all flesh) vnto whome I am afraide to commend my name, as a brocher of newe doctrines: therefore I haue beene carefull to sette down nothing, but that which will abide the triall of all whosoeuer, and abide firme against all. If any of my godly, reuered, & lerned brethren of the church

of

55

of Englande, will doe me the fauour (which thing I earnestly request at their hands) in short syllogisms to communicate with me in writing, their reasons for the ministerie of these ignorant men. I shal haue cause to praise the Lorde, & thanke them for their loue towards me in this point also, & the easier satisfie whatsoeuer may be said on the behalfe of this vnlerned (I would vnknown) ministerie. I greatly reuerence many godly and lerned, that are contrary minded vnto me in this cause; I know my yeares to be subiect vnto that bold temeritie, hoat, and heady rashnes, which blessed Timothie in consideration of his age, was bidden by the Apostle to auoid: notwithstanding, in regard of the cause (in humility before the Lord, his elect Aungels and children be it spoken) I may iustly (far be it I should) contemne & despise, almost not vouchsafe to answere, whatsoeuer man or Aungel, can bring against it. And I dare arrest and attaint of high treason, against the Maiestie of the highest, all those, both men and Aungels, who either defend the communicating with them lawful; communicate with them, or tollerate them as ministers vnder their gouernment.

IOHN PENRI.

TO THE READER.

I haue read Master D. Somes booke, the reasons he vseth in the questions of the dumbe ministerie, and communicating with them, I had answered (as you may see in this booke) before he had written. The man I reuerence, as a godly & a learned man. The weaknes of his reasons, shalbe shewed at large Godwilling. Page 42. line 24. read made and full, page 52. line 22. read axiom. Beare with the rest of the faultes.

An Exhortation continued

Courtesy British Library London

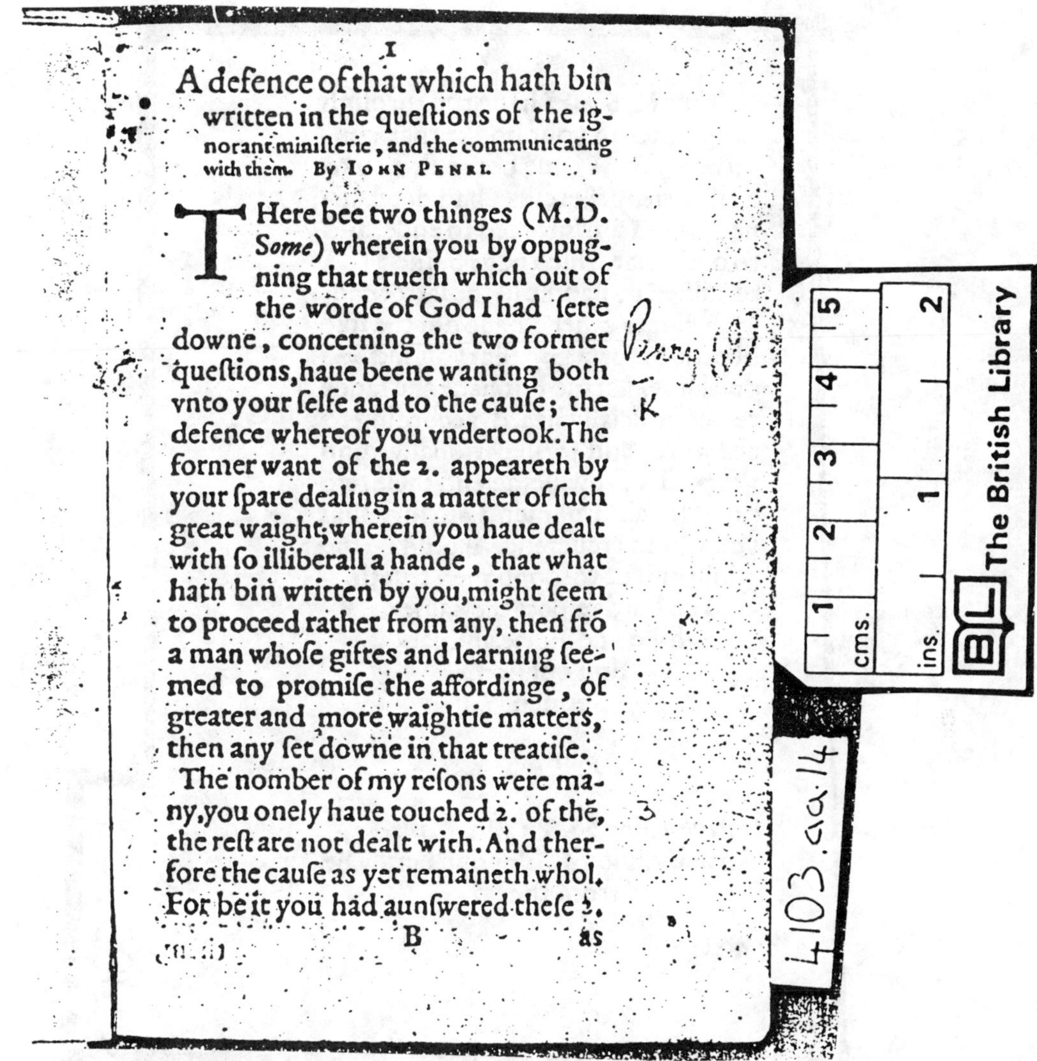

1

A defence of that which hath bin written in the questions of the ignorant ministerie, and the communicating with them. By Iohn Penri.

THere bee two thinges (M. D. *Some*) wherein you by oppugning that trueth which out of the worde of God I had sette downe, concerning the two former questions, haue beene wanting both vnto your selfe and to the cause; the defence whereof you vndertook. The former want of the 2. appeareth by your spare dealing in a matter of such great waight; wherein you haue dealt with so illiberall a hande, that what hath bin written by you, might seem to proceed rather from any, then frō a man whose giftes and learning seemed to promise the affordinge, of greater and more waightie matters, then any set downe in that treatise.

The nomber of my resons were many, you onely haue touched 2. of thē, the rest are not dealt with. And therfore the cause as yet remaineth whol. For be it you had aunswered these 2.

B as

Courtesy British Library London

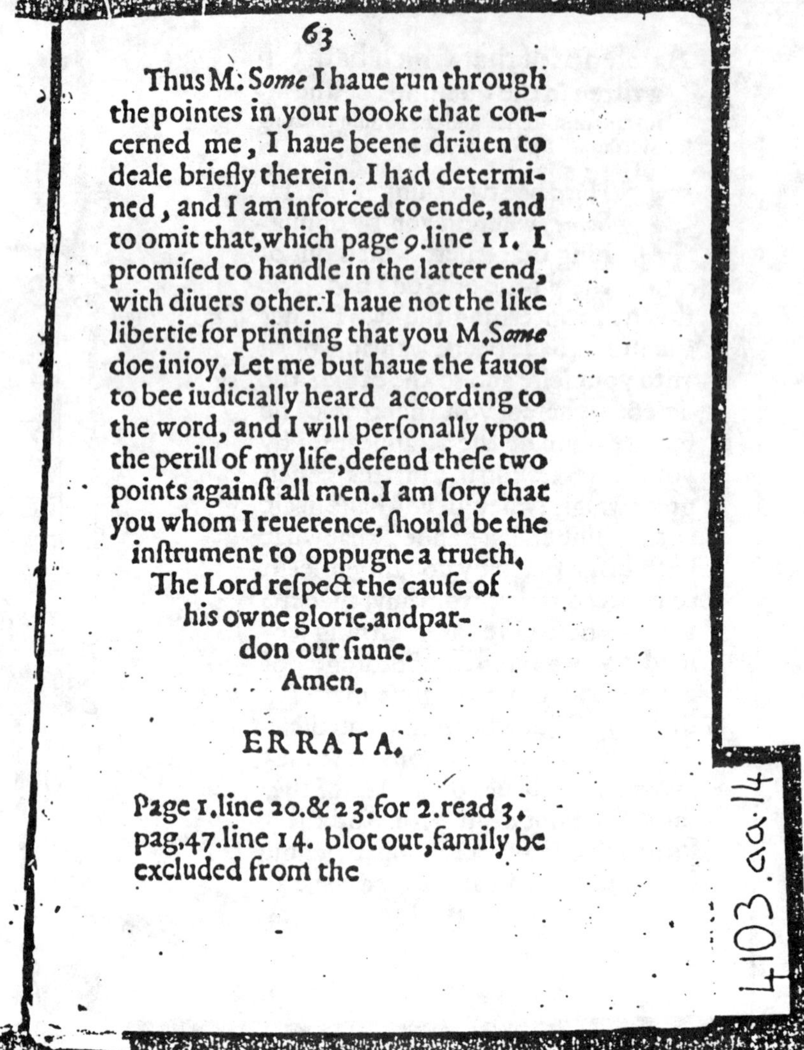

63

Thus M: *Some* I haue run through the pointes in your booke that concerned me, I haue beene driuen to deale briefly therein. I had determined, and I am inforced to ende, and to omit that, which page 9. line 11. I promised to handle in the latter end, with diuers other. I haue not the like libertie for printing that you M. *Some* doe inioy. Let me but haue the fauor to bee iudicially heard according to the word, and I will personally vpon the perill of my life, defend these two points against all men. I am sory that you whom I reuerence, should be the instrument to oppugne a trueth.

The Lord respect the cause of his owne glorie, and pardon our sinne.
Amen.

ERRATA.

Page 1. line 20. & 23. for 2. read 3.
pag. 47. line 14. blot out, family be excluded from the

A defence continued

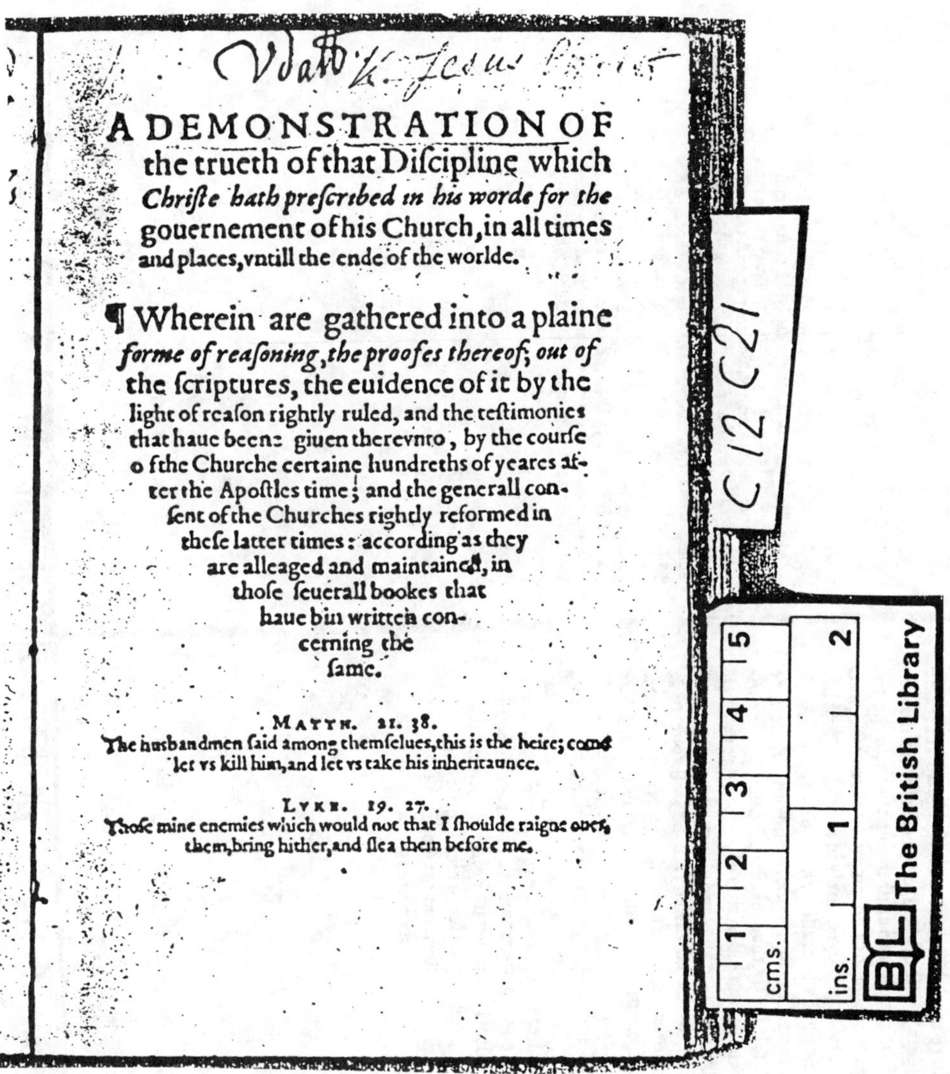

A DEMONSTRATION OF the trueth of that Discipline which *Christe hath prescribed in his worde for the* gouernement of his Church, in all times and places, vntill the ende of the worlde.

¶ Wherein are gathered into a plaine *forme of reasoning, the proofes thereof; out of* the scriptures, the euidence of it by the light of reason rightly ruled, and the testimonies that haue beene giuen therevnto, by the course of the Churche certaine hundreths of yeares after the Apostles time; and the generall consent of the Churches rightly reformed in these latter times: according as they are alleaged and maintained, in those seuerall bookes that haue bin written concerning the same.

MATTH. 21. 38.
The husbandmen said among themselues, this is the heire; come let vs kill him, and let vs take his inheritaunce.

LVKE. 19. 27.
Those mine enemies which would not that I shoulde raigne ouer, them, bring hither, and slea them before me.

Courtesy British Library London

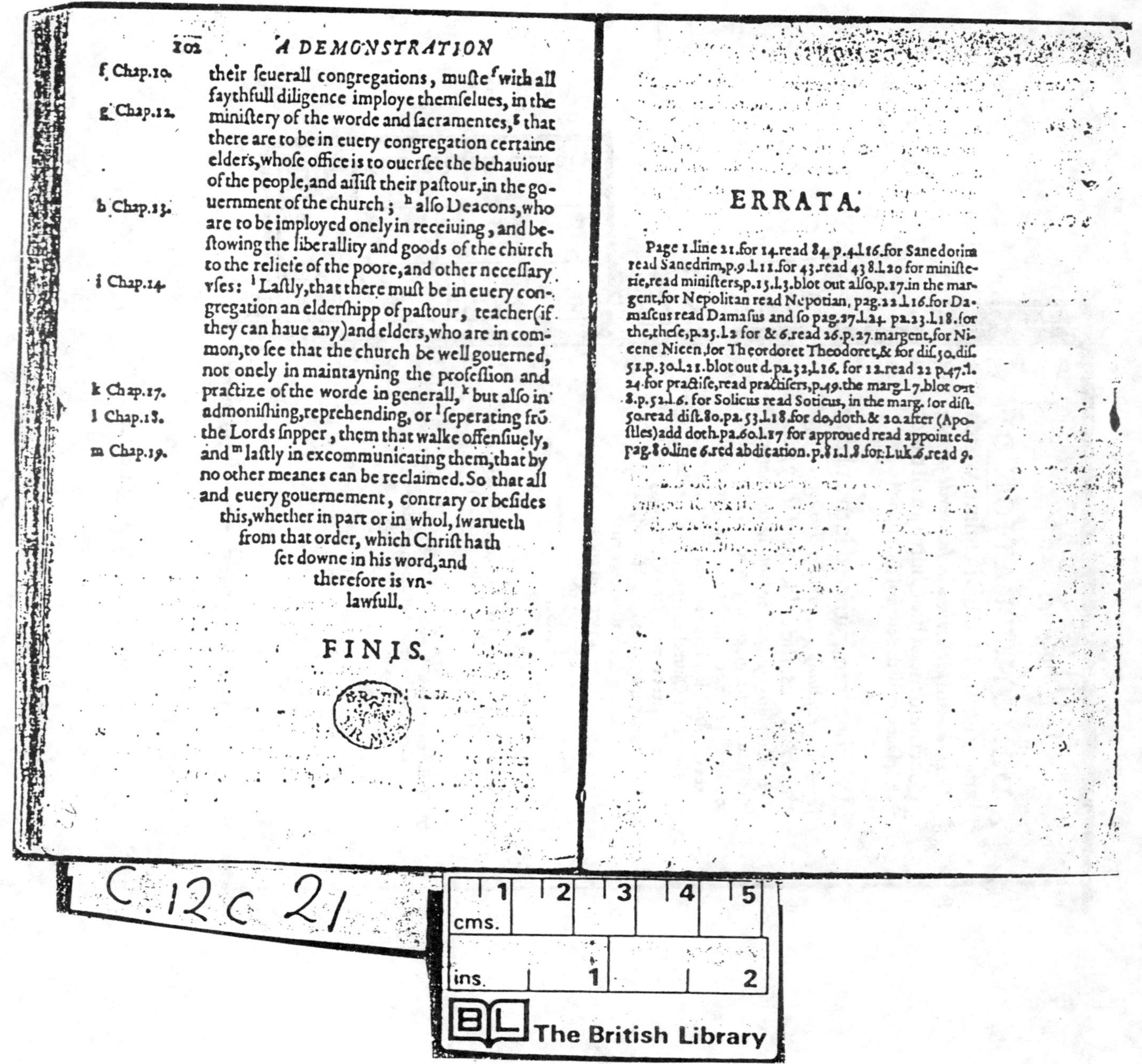

102 A DEMONSTRATION

f Chap.10. their seuerall congregations, muste [f] with all faythfull diligence imploye themselues, in the
g Chap.12. ministery of the worde and sacramentes, [g] that there are to be in euery congregation certaine elders, whose office is to ouersee the behauiour of the people, and assist their pastour, in the gouernment of the church;
h Chap.13. [h] also Deacons, who are to be imployed onely in receiuing, and bestowing the liberallity and goods of the church to the reliefe of the poore, and other necessary
i Chap.14. vses: [i] Lastly, that there must be in euery congregation an eldershipp of pastour, teacher (if they can haue any) and elders, who are in common, to see that the church be well gouerned, not onely in maintayning the profession and
k Chap.17. practize of the worde in generall, [k] but also in
l Chap.18. admonishing, reprehending, or [l] seperating frõ the Lords supper, them that walke offensiuely,
m Chap.19. and [m] lastly in excommunicating them, that by no other meanes can be reclaimed. So that all and euery gouernement, contrary or besides this, whether in part or in whol, swarueth from that order, which Christ hath set downe in his word, and therefore is vnlawfull.

FINIS.

ERRATA.

Page 1. line 21. for 14. read 84. p. 4. l. 16. for Sanedorim read Sanedrim, p. 9. l. 11. for 43. read 438. l. 20 for ministerie, read ministers, p. 15. l. 3. blot out also, p. 17. in the margent, for Nepolitan read Nepotian, pag. 22. l. 16. for Damascus read Damasus and so pag. 27. l. 24. pa. 23. l. 18. for the, these, p. 25. l. 2 for & 6. read 26. p. 27. margent, for Nicene Nicen, for Theordoret Theodoret, & for dist. 50. dist. 51. p. 30. l. 21. blot out d. pa. 32, l. 16. for 12. read 22 p. 47. l. 24. for practise, read practisers, p. 49. the marg. l. 7. blot out 8. p. 52. l. 6. for Solicus read Soticus, in the marg. for dist. 50. read dist. 80. pa. 53. l. 18. for do, doth. & 20. after (Apostles) add doth. pa. 60. l. 17 for approued read appointed. pag. 80. line 6. red abdication. p. 81. l. 8. for Luk. 6. read 9.

A Demonstration continued

Courtesy British Library London

Oh read ouer D. John Bridges/ for it is worthy worke:

Or an epitome of the

fyrste Booke/ of that right worshipfull vo-
lume/ written against the Puritanes/ in the defence of
the noble cleargie/ by as worshipfull a priest/ John Bridges/
Presbyter/ Priest or elder/ doctor of Diuillitie/ and Deane of
Sarum. Wherein the arguments of the puritans are
wisely preuented/ that when they come to answere M. Doctor/ they must needes
say some thing that hath
bene spoken.

Compiled for the behoofe and ouerthrow of
the unpreaching Parsons/ Fyckers/ and Currats/
that haue lernt their Catechismes/ and are past grace:
By the reuerend and worthie Martin Marprelat
gentleman/ and dedicated by a second Epistle
to the Terrible Priests.

In this Epitome/ the foresaide Fickers/ &c. are very in-
sufficiently furnished/ with notable inabilitie of most vin-
cible reasons/ to answere the cauill
of the puritanes.

And lest M. Doctor should thinke that no man can write without sence but his selfe/ the senceles titles of the seueral pages/
and the handling of the matter throughout the Epitome/
shewe plainely/ that beetleheaded ignoraunce/ must not liue
and die with him alone.

Printed on the other hand of some of the Priests.

Title page of the *Epitome*, the second Marprelate book, printed in November 1588 by Robert Waldegrave, at Fawsley, Northamptonshire. The first eleven lines are identical with those of the first Marprelate book, the *Epistle*, of October 1588.

Courtesy Huntington Library California and Leland H. Carlson

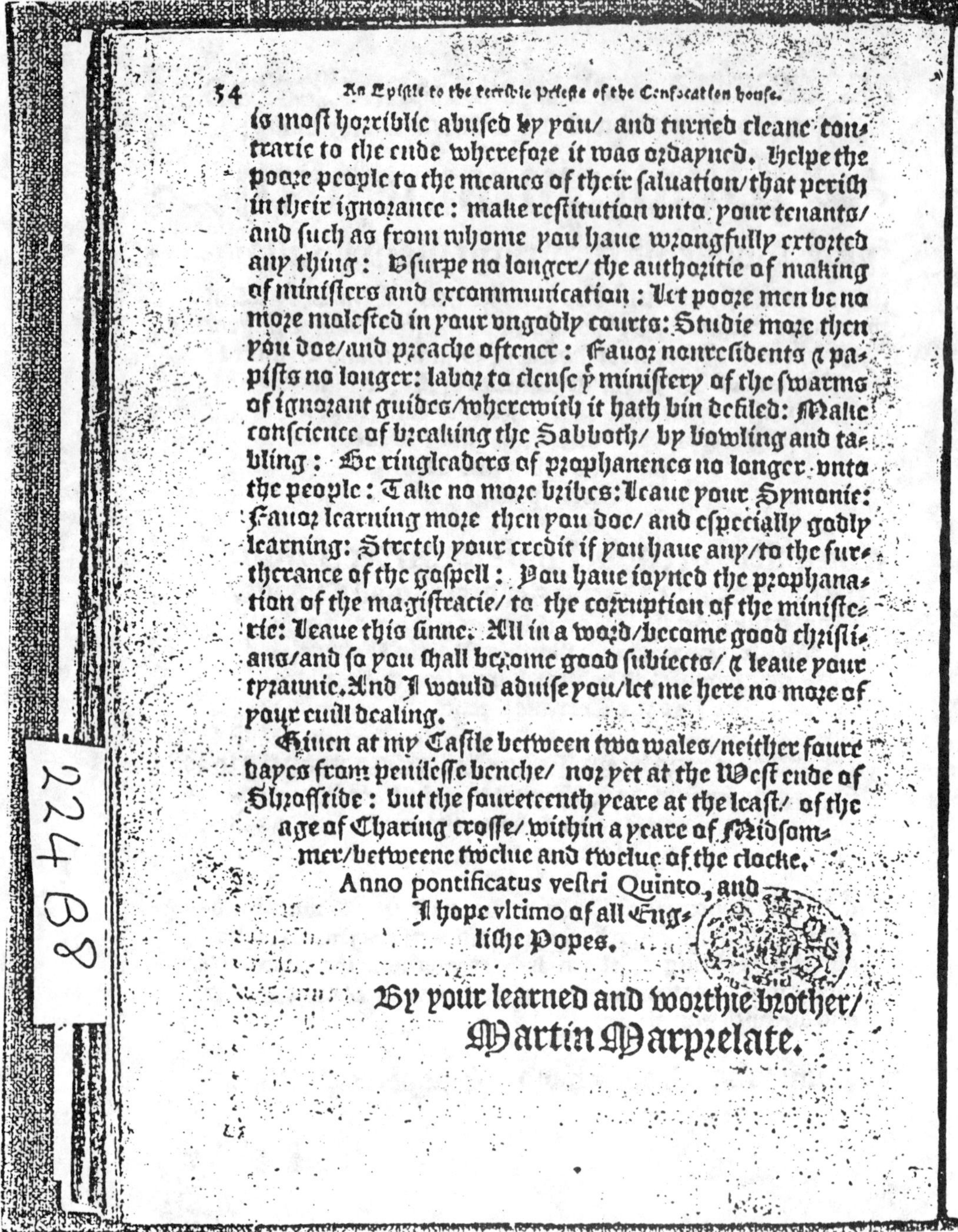

is most horrible abused by you/ and turned cleane contrarie to the ende wherefore it was ordayned. Helpe the poore people to the meanes of their saluation/that perish in their ignorance: make restitution vnto your tenants/ and such as from whome you haue wrongfully extorted any thing: Vsurpe no longer/ the authoritie of making of ministers and excommunication: Let poore men be no more molested in your vngodly courts: Studie more then you doe/and preache oftener: Fauor nonresidents & papists no longer: labor to clense ẏ ministery of the swarms of ignorant guides/wherewith it hath bin defiled: Make conscience of breaking the Sabboth/ by bowling and tabling: Be ringleaders of prophanenes no longer vnto the people: Take no more bribes: Leaue your Symonie: Fauor learning more then you doe/ and especially godly learning: Stretch your credit if you haue any/to the furtherance of the gospell: You haue ioyned the prophanation of the magistracie/ to the corruption of the ministerie: Leaue this sinne. All in a word/become good christians/and so you shall become good subiects/ & leaue your tyrannie. And I would aduise you/let me here no more of your euill dealing.

Giuen at my Castle between two wales/neither foure dayes from penilesse benche/ nor yet at the West ende of Shrafftide: but the fourteenth yeare at the least/ of the age of Charing crosse/ within a yeare of Midsommer/betweene twelue and twelue of the clocke.

Anno pontificatus vestri Quinto, and I hope vltimo of all Englishe Popes.

By your learned and worthie brother/
Martin Marprelate.

Epitome (continued)

Courtesy British Library London

Hay any worke for Cooper:

Or a briefe Pistle directed by waye of an hublication to the reverende Byshopps, counselling them, if they will needs be barrelled up, for feare of smelling in the nostrels of her Maiestie & the State, that they would use the aduise of reuerend Martin, for the prouiding of their Cooper. Because the reuerend T. C. (by which misticall letters is understood, eyther the bounsing Parson of Eastmeane, or Tom Coakes his Chaplaine) to bee an unskilfull and a deceytfull tubtrimmer.

Wherein worthy Martin quits himselfe like a man I warrant you, in the modest defence of his selfe and his learned Pistles, and makes the Coopers hoopes to flye off, and the Bishops Tubs to leake out of all crye.

Penned and compiled by Martin the Metropolitane.

(John Ap Henry, or Penry)

Printed in Europe, not farre from some of the Bounsing Priestes.

This Book is wrote against, An Admonition by T. C. [Tho: Cooper] printed an: 1589. C.l. [illegible]

Courtesy British Library London

THESES MARTINIANAE:

That is,

CERTAINE DEMONSTRATIVE Conclusions, sette downe and collected (as it should seeme) by that famous and renowmed Clarke, the reuerend Martin Marprelate the great: seruing as a manifest and sufficient confutation of al that euer the Colledge of Catercaps with their whole band of Clergie-priests, haue, or canbring for the defence of their ambitious and Antichristian Prelacie.

PVBLISHED AND SET FOORTH *as an after-birth of the noble Gentleman himselfe, by a prety stripling of his,* MARTIN IVNIOR, *and dedicated by him to his good neame and nuncka, Maister* Iohn Kankerbury: *How the yongman came by them, the Reader shall vnderstande sufficiently in the Epilogue. In the meane time, whosoeuer can bring mee acquainted with my father, Ile bee bounde hee shall not loose his Labour.*

Printed by the assignes of Martin Iunior, without any priuiledge of the Catercaps.

Courtesy British Library London

MARTIN IVNIORS EPILOGVE.

the Lorde in one day is able to bring more shame vpon thee, and that in this life, then he hath heaped blessings vpon thee now for the space of thirtie yeeres and vpward. But when I doe consider thy preheminence and promotion, I do sensiblie acknowledge it to be ioyned with a rare cursse of God, euen such a cursse as very fewe (I will not say none) in Gods Church doe sustaine. And that is thy wicked and Antichristian Prelacie. The consideration of which popedome of thine maketh me thinke, that thy other place in the ciuill magistracie, being in it selfe a godly and a lawfull calling, is so become infectious, that it will be thy bane, both in this life, and in the life to come. And I am almost fully perswaded, that, that archbishopricke of thine, together with thy practises therein, shew verely, that the Lorde hath no part nor portion in that miserabble, and desperat caytiffe wicked Iohn Whitgift, the Pope of Lambehith. Leaue therefore both thy popedome, and thy vngodly proceedings, or looke for a fearefull ende.

My second and last aduise is this in a word. Suffer no more of these haggling and profane pamphlets to be published against Martin, and in defence of thy hierarchie. Otherwise thou shalt but commend thy follie and ignorance vnto the world to be notorious. Mar-martin, Leonard Wright, Fregneuile, Dick Bancroft, Tom Blan.o Bedford, Kemp, Vnderhil, serue thee for no other vse, but to worke thy ruine, and to bewray their owne shame, & miserable ignorance. Thus far of these matters.

And mee thinkes you see, nunckle Canturburie, that though I bee but young, yet I beginne pretily

MARTIN IVNIORS EPILOGVE.

pretily well to followe my fathers steppes; for I promise you, I am deceiued, vnlesse I haue a prety smattering gift in this Pistle-making, and I feare a while I shall take a pride in it. I pray you, if you can, now I haue shewed you my minde, that you woulde be a meanes, that my vather, or my brethren be not offended with me for my presuming this of mine owne head. I did all of a good meaning, to saue my fathers papers; and it would haue pitied your heart to see, how the poore papers were raine and weather-beaten, euen truely in such a sort, as they coulde scant bee read to bee printed. There was neuer a drie threede in them. These sea-iourneys are pittifull I perceiue. One thing me thinkes my father should like in me, and that is, my modestie, for I haue not presumed, to publishe mine in as large a print or volume as my father doth his. Nay, I thinke it well, if I can driblo out a Pistle in octauo nowe and then. Farewell, good nuncle, and pay this bearer for the cariage. Iuly 22. 1589. With as great speede as I might.

Your worships nephew

MARTIN IVNIOR.

THESES MARTINIANAE (contd)

F BIRCHLEY HALL PRESS, BILLINGE, near WIGAN, LANCASHIRE

from 1604?: 1616?: Raided c. 1621 (Catholic)

The population figures for 1717 for the village of Billinge show a total of 198 families of which 174 were listed as Protestants, 10 as Papists and 14 as Dissenters. From this and other evidence in the Wigan Reference Library it seems likely that during the century preceding this date the area was traditionally Protestant.

Yet it was in this seemingly dangerous environment that the Anderton family of Birchley Hall clandestinely transformed their home into a centre of Catholic resistance, and established there a printing press.

At least 16 books appear to have been printed on this secret press and one Wigan authority estimates the total number of books and pamphlets printed there to have been over 60.

The Anderton family's first book The Apologie of the Roman Church,* a small quarto of about 190 pages, is dated 1604 without place of printing, and Joseph Gillow, author of Bibliographical Dictionary of the English Catholics, considers this to have been printed in Lancashire, either at Lostock Hall near Bolton, or at Birchley Hall, near Wigan. So also does Arthur J Hawkes in his Bibliography of Lancashire Printed Books before 1800, published in 1925. The book "bears some evidence of amateur production and limited types", and contains a "handsome and curious wood-cut initial."

Notwithstanding the authorative view of the Dictionary of National Biography, James Anderton is NOT now thought to have been involved in authorship. He was, however, instrumental in the secret printing activity, being the sympathetic owner of both Lostock Hall and Birchley Hall at that date, chiefly residing at the latter from 1581 until his death in 1613 as will shortly be described.

* See Title-page Facsimile

Directly concerned with authorship of the series of Birchley Hall books was James' first cousin Lawrence Anderton S.J. (Society of Jesus) (1575-1643), a graduate of Christ's College, Cambridge, "where he displayed much brilliance and great eloquence". It is thought that the 1604 *Apologie* was written by Lawrence while based in Lancashire. In the following year, 1605, he went to Rome and there entered the Jesuit Society. A supposition is that before returning to England he bought types and printing apparatus from one of the centres of English Catholic printing on the continent such as Douai, Rouen or St Omer. Certainly the type used for the Birchley Hall books is very similar, causing earlier bibliographers to have ascribed Birchley Hall books to those towns. Most of the English books printed at Douai, Rouen and St Omer have however this fact stated in the imprint, together with the name of the printer; and several of Laurence Anderton's own books bear such imprints. If a book was printed abroad there was no need for secrecy, whereas there was considerable need if it was actually printed in England. It is therefore surely significant that, without exception, the books listed as emanating from the Birchley Hall Press bear no place of printing nor name of printer.

In 1608 the second Birchley book appeared, *The Protestants Apologie,**the first one to be written by the fictitious "John Brereley Priest". This was a much enlarged edition of the 1604 publication, the latter comprising 750 pages to the former's 193 pages, both being small quarto in page size. A substantial book indeed for a secret press, with fair typography but "abounding in misprints - the common feature of Birchley books." Special features of this 1608 book are "certain tail ornaments cut slightly lopsided, also two peculiar wood-cut capitals "HH" with a slight blemish".

* See Facsimiles

In 1613 James Anderton's "goods and books" were seized. This information is contained in a 20th November letter from Sir Julius Caesar, Chancellor of the Exchequer, referring to "the safe custody of the goods and books of one Anderton, a recusant in Lancashire, deceased." A supplementary letter by the same writer gives an inventory of the confiscated books, and describes the short title of each book _in the plural_, as for example, "Vauxs Catechisms", whereas the title of the actual 1605 book is _A Catechisme_ by Lawrence Vaux. This implies the seizure of a number of copies of the same book for each item listed, as one would find either in a distribution depot or in a printing establishment.

A second seizure of Anderton property occurred in about 1621 and this time the contemporary evidence proves the existence of a printing press. An anti-Roman Catholic tract _The Foot out of a snare_ published in 1624 states: "There was a printing house suppressed about three years since [i.e. in 1621] in Lancashire where all Brerely his works, with many other Popish pamphlets were printed". (See DNB, Anderton, James p 396). The author of this work, John Gee (See DNB p 986) was for a time a convert to the Church of Rome, who then in 1624 printed detailed evidence against his recent co-religionists. Further he had Lancashire connections and in 1622 he was beneficed at Newton, near Winwick, only about ten miles distant from Birchley Hall, so that his evidence as to the existence and suppression of the press is likely to be reliable.

The "Anderton ---- deceased", in the quotation above-mentioned, was James Anderton who died at Lostock in September 1613. It was not cousin Lawrence, but James' younger brother Roger Anderton who now succeeded to the Lostock Hall and Birchley Hall properties. Roger too is regarded as a key figure in the Birchley Hall publications and he may well have superintended the printing, though is thought to have played a part in authorship as well. A list comprising 25 Birchley books (some of them never located) is extant, designated as "the workes of my Uncle Rog[er] and [erton] which was sent me by his son C[hristopher] Anderton AD 1647". Roger Anderton had died at Birchley in 1641 and this list was made six years later by his son, then aged 40, apparently from an original compiled by Roger Anderton himself, for it is headed "original coppy owne hand". Many of the books thus listed are known to have been written or translated by others, so the only way in which this list of 25 books can be said to be "the workes" of Roger Anderton is that he printed them.

The dates (where known) of the books clearly designated to the Birchley Hall secret press are 1604, 1608, 1613, (the date of the book-stock seizure); 1615, 1619, 1620; (seizure of the press c.1621); 1623; 1624; 1630; 1633; 1635; 1640; and 1642. Although firm evidence is lacking, it is therefore probable that the Birchley press may have been re-established after its seizure in about 1621, because the above-mentioned list of "the workes" of Roger Anderton include Birchley books dated 1623 to 1642 inclusive as detailed in the above list of dates. Lawrence Anderton also continued to be zealous for Roman Catholicism after the seizure of the press (probably in 1621) because in that same year he was appointed from Rome to the position of Superior of the Lancashire Provinces, and he certainly continued to write catholic works thereafter.

"When the [Birchley Hall Press] was discontinued cannot be stated with any exactness, nor what became of it afterwards. So far no book has come to light which has any pretension to Birchley printing later than 1642 [the year of Charles I's journeying through England with his royal printer with England's Civil War beginning that same autumn]. It seems likely therefore that the death of Roger Anderton in 1641 marked the beginning of the end of this secret press, and that its final close came with the death of Lawrence Anderton in April 1643".

In concluding this narrative of intriguing but elusive illegal printing, mention should be made of visual similarities between Edmund Campion's "Ten Reasons" printed at Stonor Park near Oxford in 1581 (as above described) and the first Anderton book, Apologie of 1604. The Jesuit emblem in an oval on the two title-pages is somewhat similar, as also is their general lay-out. Lawrence Anderton was certainly acquainted with Campion's Ten Reasons because an entirely original translation of it was printed at Birchley in 1632. So it seems possible that Campion's 1581 book formed Anderton's styling model in 1604. (These two IHS designs appear on the front cover).

I am greatly indebted, in respect of information in the early part of this section, to Wigan Reference Library, who sent me useful exerpts from their copy of Lancashire Printed Books before 1800 by A J Hawkes (1925).

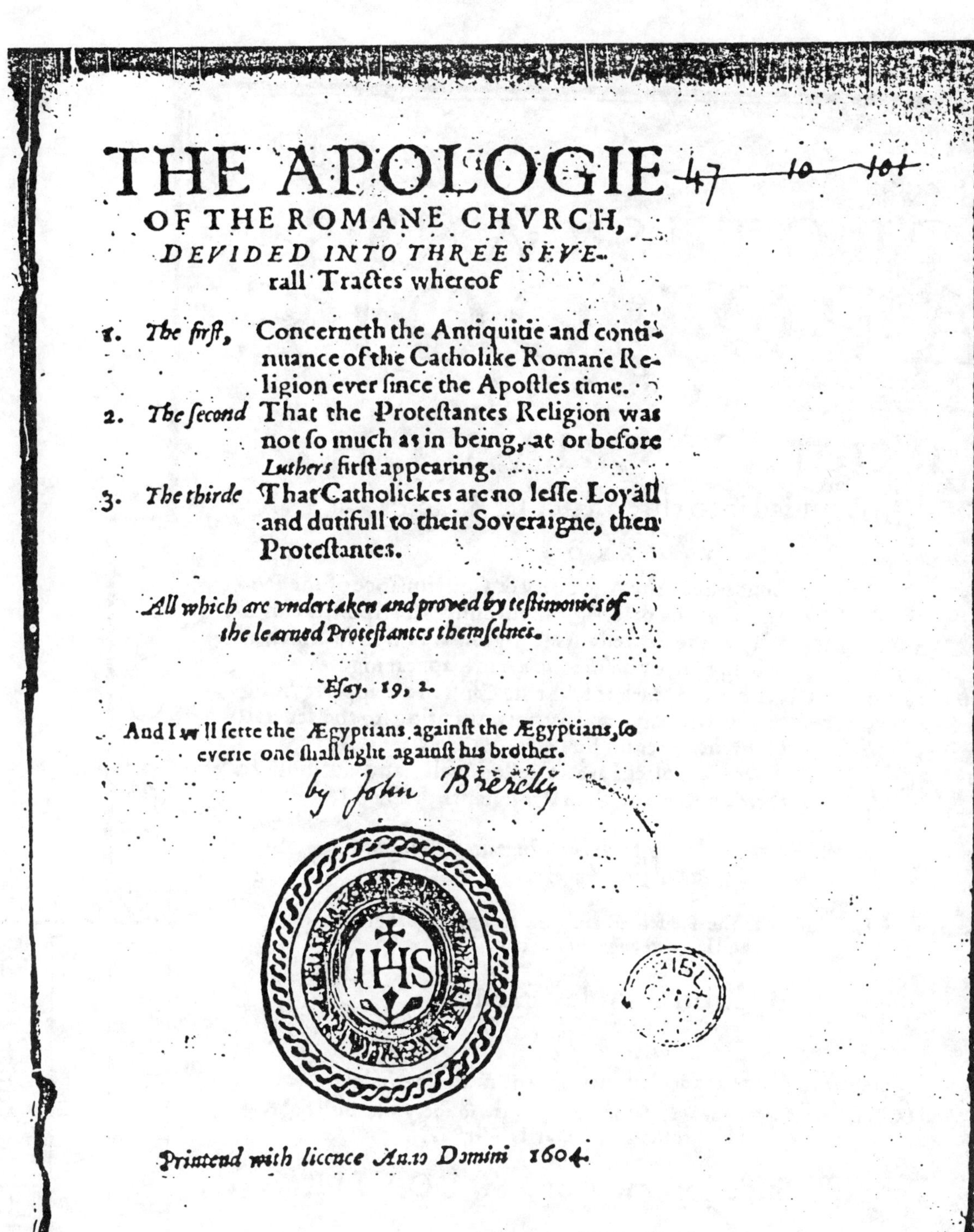
THE APOLOGIE
OF THE ROMANE CHVRCH,
DEVIDED INTO THREE SEVE-
rall Tractes whereof

1. *The first,* Concerneth the Antiquitie and continuance of the Catholike Romane Religion ever since the Apostles time.
2. *The second* That the Protestantes Religion was not so much as in being, at or before *Luthers* first appearing.
3. *The thirde* That Catholickes are no lesse Loyall and dutifull to their Soveraigne, then Protestantes.

All which are undertaken and proved by testimonies of the learned Protestantes themselves.

Esay. 19, 2.

And I will sette the Ægyptians against the Ægyptians, so everie one shall fight against his brother.

by John Brerely

IHS

Printed with licence Anno Domini 1604.

Courtesy British Library London

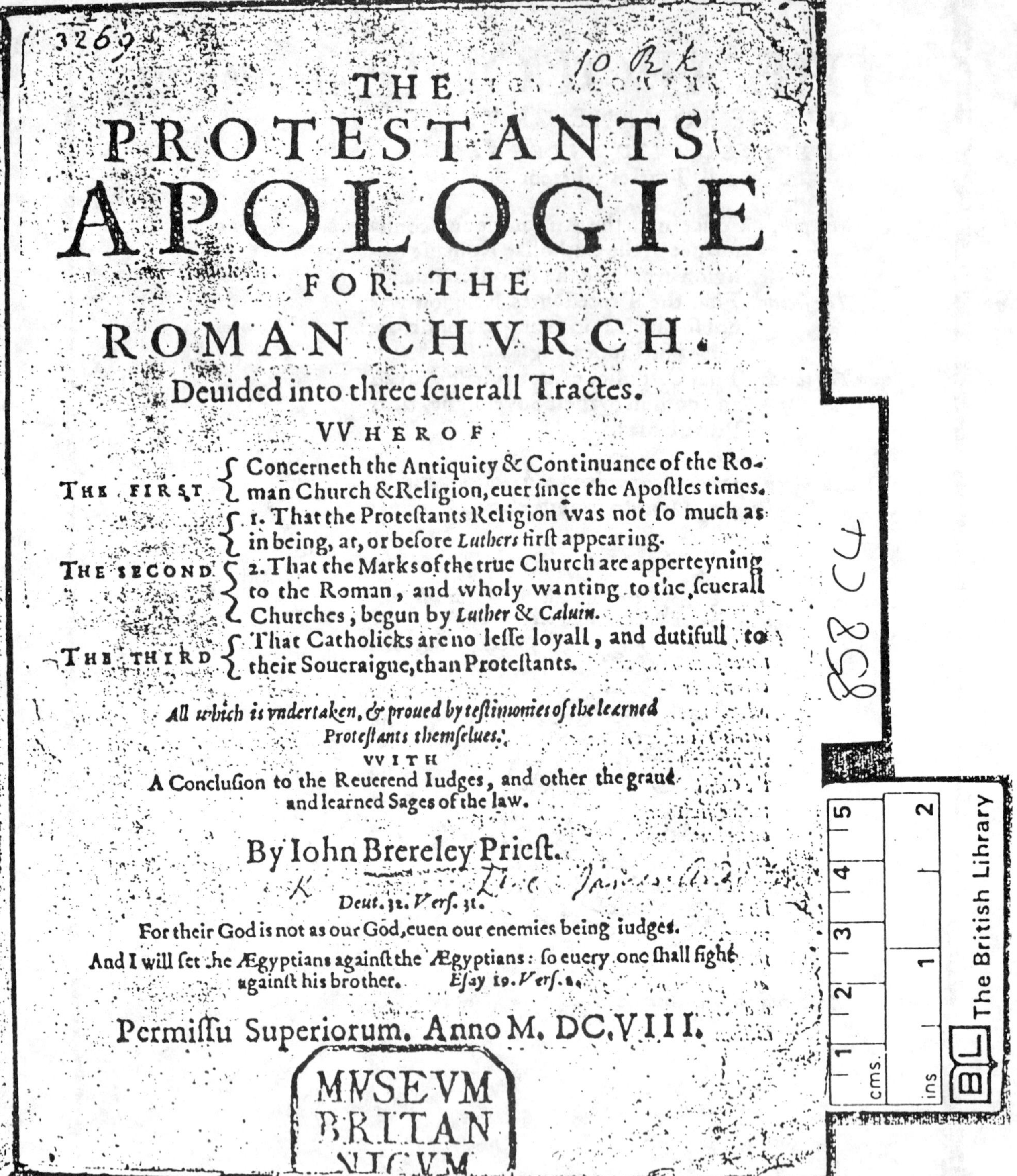

THE
PROTESTANTS
APOLOGIE
FOR THE
ROMAN CHVRCH.
Deuided into three seuerall Tractes.
VVHEROF

THE FIRST { Concerneth the Antiquity & Continuance of the Roman Church & Religion, euer since the Apostles times.

THE SECOND { 1. That the Protestants Religion was not so much as in being, at, or before *Luthers* first appearing. 2. That the Marks of the true Church are apperteyning to the Roman, and wholy wanting to the seuerall Churches, begun by *Luther* & *Caluin*.

THE THIRD { That Catholicks are no lesse loyall, and dutifull to their Soueraigne, than Protestants.

All which is vndertaken, & proued by testimonies of the learned Protestants themselues.
VVITH
A Conclusion to the Reuerend Iudges, and other the graue and learned Sages of the law.

By Iohn Brereley Priest.

Deut. 32. Vers. 31.
For their God is not as our God, euen our enemies being iudges.
And I will set the Ægyptians against the Ægyptians: so euery one shall fight against his brother. *Esay 19. Vers. 2.*

Permissu Superiorum. Anno M. DC.VIII.

Courtesy British Library London

therein vsed by vs at this day are truly such as are lawfully and lineally descended from our noblest progenitors the auncient Fathers.

But seeing that which is offered in this sacrifice, is that which is most precious, to wit, the sacred humanity of Christ our Sauiour, here shal you likewise see it most euidently euicted, and that by al argumentes conuincing true faith and religion, that in the sacred Eucharist Christes body is not onely by way of figure or remembrance, according to *Swinglius*, neither only by faith and spiritually, as *Caluin*, neither conioyned with bread as *Luther*, but that the bread and wine according to their substances and natures, are truly and really chaunged into the body and bloud of Christ, then which, what more grateful can be offered to God, what more honourable can be left to his Church, or what more profitably and comfortable can be bestowed vpon the soules of the iust.

This then, most noble Prince, being the heauenly foode which I humbly presume to offer vnto your excellencies most sweetest taist and contemplation, I shal rest in hope that the dilicacy thereof wil so strongly possesse your senses with liking and admiration, as that therby pardon wil be pleaded for so daring attempt, as the presuming to write to so great a soueraigne. In confidence whereof I wil remaine.

Your highnes humble seruant, and deuoted Orator

IOHN BRERELEY.

The Protestants Apologie..... (continued)

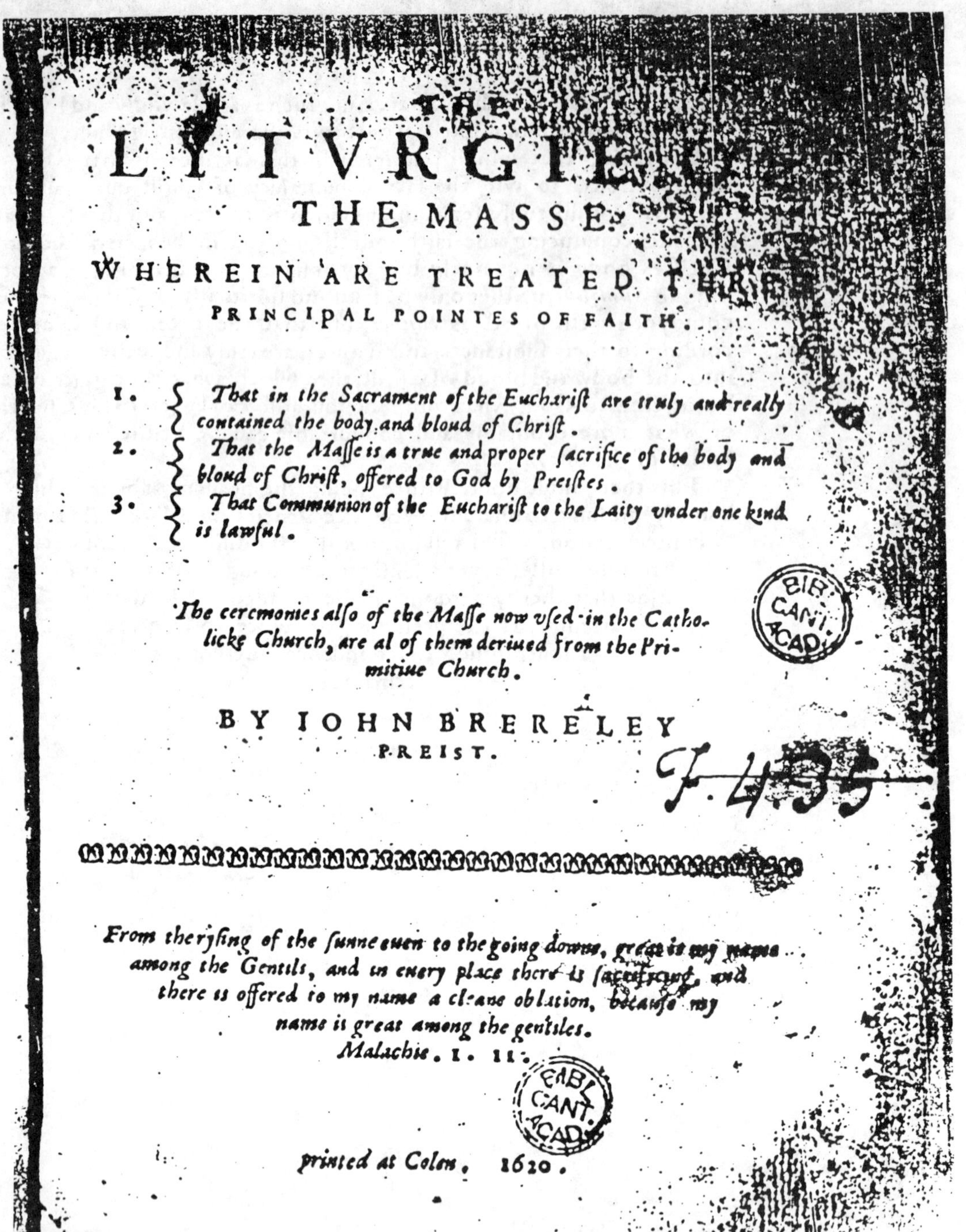

THE

LYTVRGIE OF

THE MASSE:

WHEREIN ARE TREATED THREE
PRINCIPAL POINTES OF FAITH.

1. That in the Sacrament of the Eucharist are truly and really contained the body and bloud of Christ.
2. That the Masse is a true and proper sacrifice of the body and bloud of Christ, offered to God by Preistes.
3. That Communion of the Eucharist to the Laity vnder one kind is lawful.

The ceremonies also of the Masse now vsed in the Catholicke Church, are al of them deriued from the Primitiue Church.

BY IOHN BRERELEY
PREIST.

From the rysing of the sunne euen to the going downe, great is my name among the Gentils, and in euery place there is sacrificing, and there is offered to my name a cleane oblation, because my name is great among the gentiles.
Malachie. 1. 11.

printed at Colen. 1620.

Courtesy British Library London

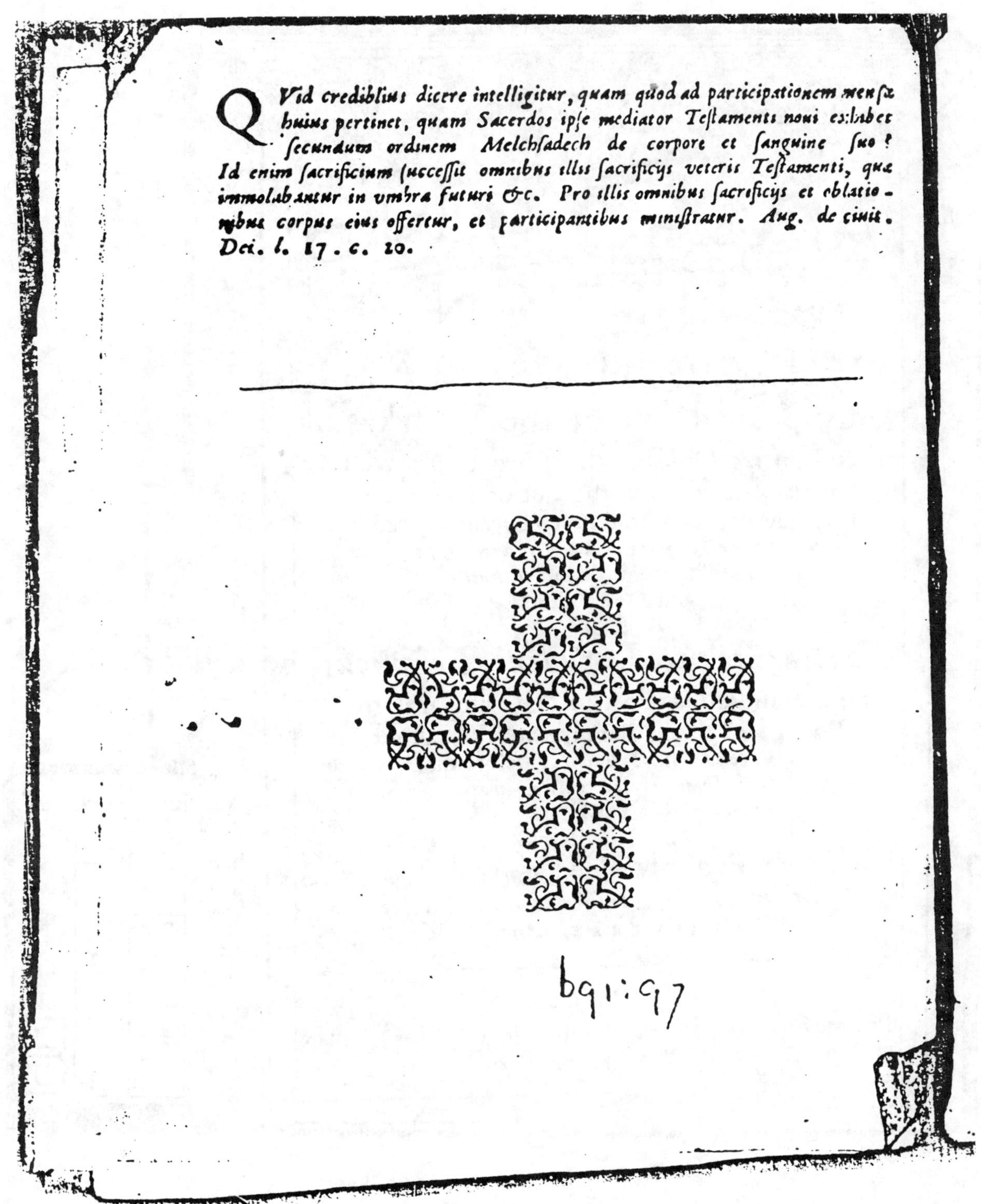

QVid credibilius dicere intelligitur, quam quod ad participationem menſæ huius pertinet, quam Sacerdos ipſe mediator Teſtamenti noui exhibet ſecundum ordinem Melchiſedech de corpore et ſanguine ſuo? Id enim ſacrificium ſucceſſit omnibus illis ſacrificijs veteris Teſtamenti, quæ immolabantur in vmbra futuri &c. Pro illis omnibus ſacrificijs et oblationibus corpus eius offertur, et participantibus miniſtratur. Aug. de ciuit. Dei. l. 17. c. 20.

The Litvrgie of the Masse (continued)

Courtesy British Library London

SAINT
AVSTINS
RELIGION:

VVHEREIN IS MANIfestly proued out of the Workes of that *Learned Father*, who liued neere twelue hundred yeares before the time of LVTHER; that he dissented from *Poperie*, and agreed with the *Religion* of the PROTESTANTS, in all the maine Poynts of *Faith* and DOCTRINE:

Contrary to that Impudent, Erronious, and Slanderous Position of the bragging PAPISTS of our Times, who falsly affirme, *Wee* had no *Religion*, before the *Times of* LVTHER *and* CALVINE.

Whereunto is newly added, *Saint* AVSTINS SVMMES; *in Answer to Mr.* IOHN BREERELY, *Priest.*

LONDON.
Printed for *Iohn Marriot*, and are to be sold at his Shop in St. *Dunstane's* Churchyard in *Fleet-street*. 162[illegible]

"IN ANSWER TO MR IOHN BREERELEY, PRIEST"

Courtesy British Library London

OTHER POSSIBLE LOCATIONS

"Happy are the Secret Presses which have no History" (which is of course a paraphrase of the well-known if slightly negative saying "Happy is the Country which has no History).

All of the five different secret presses summarised above were raided or seized by governmental or ecclesiastical agents, and much detail is known about them for those very reasons. Clearly the authorities pursued their searches with alacrity and vigour, for the first two presses which I have described remained undetected for less than two years, and the press in the Welsh cave has left no extant title known to me. The Marprelate Press despite moving to six locations was twice raided yet still struggled on. Only the Catholic Birchley Hall press near Wigan survived for a measure of years, until it in turn was raided in 1621. Of the five described it enjoyed the greatest longevity.

So were there other secret provincial presses which remained hidden and therefore unknown to posterity? Are there any contemporary "printing whispers in the wind" as to the possible existence of other country presses which succeeded in keeping their secret. For this same span of decades, I can suggest two, namely:

(a) "THACKWELL PRINTED POPISH BOOKS IN WALES"

In the Marprelate tract _An Epitome_: _Oh read over D. John Bridges. for it is worthy worke_ (1588) a contrast is drawn between the usage Waldegrave received and that which "a printer named Thackwell, who had printed popish books in Wales had received". Marprelate writes "Thackwell is at libertie to walke where he will, and permitted to make the most of his press and letters." (p 396 _Annals of Scottish Printing_ by Dickson and Edmond).

(b) Sheila Lambert in a lecture in 1986 (see bibliography) quotes a case in about 1637 of three men who "were found printing the Psalms at HENLEY". For this offence of pirating Stock books they apparently avoided the pillory, but were sentenced to be whipped.

(c) For WORCESTER there is an unconfirmed local reference to the establishment of a local newspaper in 1690. (See my Ipswich-Worcester 'green-back' page 161.)

(d) <u>OTHER POSSIBLE CIVIL WAR MOBILE PRESSES:</u> I have a note that Falconer Madan, in a Preface p ix of one of this three epic <u>Oxford Books</u> writes of the possibility of another wandering press at work in February 1642/1643. (This remains for me to study further). There is also an outside possibility of a further Civil War Puritan press at Shrewsbury during the Parliamentarian occupation of that city. (See my King's Printer in York and in Shrewsbury p 28).

If anyone knows of other tenuous early "printing whispers" in Wales or in England outside London, I will be pleased to hear.

H WORKING CONDITIONS AT THE SECRET PRESSES

Examination of a sample of the works of these secret presses readily reveal evidence of shortage of type, as of a higher than average incidence of spelling errors or of printing ommissions of pagination, signature marks etc. Risk of detection would be ever-present, making it unlikely that the working press would be accommodated in the main windowed rooms. The most likely safe place would be in cellars, which might normally be kept locked. Heating would either be absent or dangerous (for smoke from a cellar fire-place might mean added risk). Likewise the work would most likely be entirely by candle-light or inadequate oil lamp.

Many of these illegal works were therefore likely to have been produced in conditions of cold and poor lighting, and if so posterity should be amazed, not at the frequency of error, but rather at the high-standard of the work produced, in conditions which would be judged intolerable by today's high standards of physical comfort.

IV

ROBERT WALDEGRAVE (c.1551 or 1554 to 1604)

(Printer of London: Who-knows-where: La Rochelle: Edinburgh and London)

For one man to be firstly a Stationers Company apprentice and London master-printer; secondly to be perilously working an on-the-run press; thirdly to print in France; fourthly to become the Scots royal printer; and finally to return ill but triumphant to London once again, may well be unique in its dramatic variety; yet this in a nutshell was the achievement of the resourceful and resilient printer, Robert Waldegrave.

Dickson & Edmond in their *Annals of Scottish Printing* towards the end of their classic work devote three chapters to Waldegrave, to his dated and to his undated works (pages 394 to 474) and they begin with his law-abiding, early career in London.

The registers of the Stationers Company record the indenture on 24-6-1568 for eight years of "Robert Walgrave the sonne of Rychard Walgrave late of blacklay in the Countye of Worcestre yeoman Deceased ----- to Wylliam greffeth Cetizen and stacioner of London ----"

If young apprentice Robert was of the usual age of fourteen, then his date of birth may be set at circa 1554. However, another record of the Stationers Company quotes him as having been a master printer eight years by March 1581, so with the known eight years apprenticeship duration and the conjectured fourteen year commencement, this would yield a birth year of 1551.

In a list of 'printers in London keping presses ---' compiled in 1583, Waldegrave is entered as having two presses (Barker the Queen's Printer having five by comparison), and Waldegrave's two presses could be of special importance in the context of the first stages of his second career as secret printer.

The Stationers Registers of books licensed to Waldegrave were numerous, with dates between the years 1578 and 1588. Perhaps therefore he worked for others for a few years after completing his apprenticeship, before, still in his twenties, setting up on his own account for a decade of open, allowed printing in the metropolis, during which time there are records of him engaging three apprentices. He was once enjoined not to print "any thing in master Seres privilege" but nothing untoward otherwise. I have, for example, inspected at the British Library London two of Waldegrave's London printings for the Kingston upon Thames preacher John Udall: a 1588 volume of three sermons Amendment of Life (B.L. ref. 4479 a 32); and another of four sermons titled The Combate betwixt Christ and the Deuill, which includes on the final verso a woodcut of Christ striking the devil on the head with a ball-and-chain, together with the initials R and W in the lower corner. (See facsimiles).

But then, as indicated at the beginning of the Marprelate account, Waldegrave's London press (singular) was seized. This occurred on 16th April 1588 and included divers copies of an unauthorised book The State of the Churche of England laid open --- (also known as Diotrephes), and believedly written by Job Throkmorton, whom Leland Carlson's book identifies as the Martin Marprelate author.

Dickson & Edmond write that Waldegrave was "kept a prisoner for twenty weeks in the White Lion". Meantime his press and types and books were brought before the Court of the Stationers Company on 13th May 1588 to be burnt, defaced and made unservicable - probably in the now peaceful enclosed garden to the left of Stationers' Hall forecourt.

It should be noted that Dictionary of Printers ..... (1557-1640), in its Waldegrave entry on pages 277-279, gives different dating of multiple imprisonment. This source states that Waldegrave was imprisoned in 1584 for printing two named books. "For this Waldegrave was thrown into the White Lyon prison in Southwark for six weeks, and again in 1585 he was imprisoned there for twenty weeks, for printing Puritan Literature. (see Hay any worke for Cooper, ed Petheram, p. 68)"

In the November 1588 Marprelate Tract Epitome, sub-titled 'Oh read ouer D. John Bridges/for it is worthy worke' the following graphic description appears:

> .....Robert Waldegrave dares not shew his face for the bloodthirstie desire you have for his life --- You know that Waldegrave's printing press and letters were taken away: his press being timber was sawen and hewed in pieces: that yron worke battered and made unserviceable; his letters melted, with cases and tools defaced (by John Woolfe, alis Machivell, Beadle of the Stacioners and most tormenting executioner of Waldegrave's goods) and he himself utterly deprived of ever printing again, having a wife and sixe small children.

If however we discount the Dictionary of Printers imprisonment dating, we learn that Robert Waldegrave's press was seized on 16th April 1588 and that he himself was imprisoned for twenty weeks, which would mean his being released in early August 1588. However, Waldegrave had no intention of being "utterly deprived of ever printing again"; instead imprisonment and violence to his means of livlihood had reinforced his anti-episcopal, puritan beliefs. Evidently one of his presses and some of his types had escaped seizure; (perhaps he had moved them to another location when he began to print Throkmorton's unauthorised Diotrophes book as a precaution against such a visitation, for his career proves him to be a most resourceful man. Alternately he may have acquired fresh apparatus).

Thus in the weeks between early August and October 1588, his secret press, in a state of readily hidden mobility, was established, as we have seen in the Marprelate narrative, at Mistress Crane's country house at East Molesey in Surrey, where two,perhaps three,clandestine works were produced including the first of the Marprelate series.

It may well have been in the lull between printing and October publishing that the secret press, well-hidden in its cart, successfully made its second perilous move, this time northwards to the Midlands, to Fawsley near Daventry at the house of Sir Richard Knightley, where Epitome the second in the Marprelate series was successfully printed and distributed in November 1588.

Soon after this the mobile, clandestine press moved for the third time, again successfully, to a house called White Friars in Coventry belonging to another puritan sympathiser John Hales. Here between December 1588 and March 1588/9, the press succeeded in producing three more works, as already described: Mineralls, John Penry's A View.... and Hay Any Worke for Cooper.

Much of this under-cover detail is known from subsequent questionings and torturings. For example Dickson & Edmond (pages 398-400) quote at length a "Secret report to Lord Burghley of the Authors of the Martin Marprelate tracts" dated 15-9-1589 (Landsowne. M5.61.Art.22). Robert Waldegrave is named as Marprelate printer in this report to Lord Burghley. An attempt had been made in our February 1589 at Northampton to arrest Henry Sharpe, Marprelate binder and distributor. It was evident that the vigilant poursuivants were closing in; Waldegrave had already suffered imprisonment for twenty weeks in the previous summer; so it is hardly surprising to learn that Waldegrave handed over the working of the secret press to John Hoskins and others named in the Marprelate descriptions and fled secretly and successfully to the continent of Europe, departing from Coventry in early April.

It is said that Waldegrave took to the continent with him the black-letter types used for printing the latest works from the Coventry press, leaving the subsequent Marprelate tracts with a smaller type range. Journeying incognito, yet carrying the bulkiness of complete alphabets of founders' type, demonstrates yet again a man of great resourcefulness and determination.

Waldegrave fled probably to Rochelle or La Rochelle, which lies on the French coast south of Nantes and north of Bordeaux. For a lead-weighted fugitive leaving Coventry in the English West Midlands, the likely transport logic would have been to try to board a small trading vessel at some port on the River Severn, say at Gloucester, or even higher up at Worcester, whence fled on the accession of Mary Tudor, 6-7-1553, another Protestant printer John Oswen, nearly 36 years earlier. (See my "green-back" The First Printers at Ipswich in 1547-1548 and at Worcester in 1549-1553.)

According to Leland Carlson's Marprelate book (p 88) Waldegrave may, between the months of April and August 1589, have worked at the printing shop of Les Haultin at La Rochelle, where he is believed to have printed "Penry's Appellation*, Throkmorton's Master Some Laid Open, and possibly his A Dialogue wherein is* plainely laid open ----".

Carlson suggests (p 159) that Waldegrave may have had the manuscript of A Dialogue when he left England about 5 April

> and that he printed it at La Rochelle in the summer of 1589. On folio A2 *recto* there is a factotum—two naked cherubs—with no enveloping border, exactly like the factotum used by Waldegrave in *Master Some Laid Open*, also printed at La Rochelle in the summer of 1589, and in *An Humble Motion*, printed by Waldegrave in May 1590 at Edinburgh. The type used in Penry's *A Viewe*, in *Martin Junior*, and *Martin Senior* was probably cast by the house of Les Haultin in La Rochelle, and was closely similar to, if not identical with, the type used to print the three La Rochelle imprints—Penry's *Appellation*, and Throkmorton's *Master Some Laid Open*, as well as *A Dialogue*.[3]
>
> 3. Penry's *Appellation* has an initial ornament letter T on page 1, which is identical with that used by the French printer in La Rochelle, Jerome Haultin. See Louis Desgraves, *Les Haultin, 1571–1623* (Geneva, 1960), p. 62. Haultin, who was in England about 1574–85, was a Huguenot and probably a close friend of Waldegrave, whose italic type in the Marprelate books was cut by Haultin. See Talbot Baines Reed and A. F. Johnson, *A History of the Old English Letter Foundries* (London, [1952]), pp. 96, 161.

* See Facsimiles

Again Carlson writes (p 88):

> Waldegrave left La Rochelle about the end of August 1589 [perhaps along the same provenly safe route] and brought to Throkmorton's house at Hasely [in Warwickshire] three or four consignments of books in September. He then departed for Edinburgh, where he arrived about November [1589]. On 13 March 1590 he received permission from 'the Lordis of the Secreit Counsell' to print The Confession of Faith Subscrived [sic] by the Kingis Majestie.

Aldis in his List of Books Printed in Scotland ---- cites as his No. 208, one book possibly printed by Waldegrave in 1589, namely: [Spiritval prepine --- (by Jas.Melville) Waldegrave, 1589, for 1598 qv.] This is STC 17816 with extant copies in the National Library of Scotland and in the British Library, London.

For the year 1590 Aldis lists no less than 14 extant imprints, all definitely or probably printed by Waldegrave and by none other in Scotland that year. This output compares with a total of but four extant titles for 1589 including the one above specified.

Three of these 1590 Waldegrave Scottish titles (all quartos) were believedly by John Penry, who had also fled north to Scotland in the previous year. The first to be listed by Aldis was Penry's Briefe discovery of the vntrvthes--*-- (Aldis 209: and STC 19603). A second was his Hvmble motion ---vnto - Privie Covnsell (Aldis 215: but not in STC'). The third of Waldegrave's Scottish 1590 imprints by Penry (Aldis 221 and STC 19612) was A Treatise Wherein is Manifestlie Proved --- * sometimes erroneously titled Reformation No Enemie (It should at least read 'enemies' - see facsimile). This as will be seen is dated 1590. It bears no imprint nor author's name, yet John Penry's indictment was based on this book (with no reference at the trial to Marprelate authorship); and it was this book which resulted in Penry being hanged at Southwark after his May 1593 conviction. (It will be noted again how serious, even dull, are the titles of John Penry's works, as compared with the jokey, catchy Marprelate titles).

* See Facsimiles

So now Robert Waldegrave reached his peak in what I regard as his fourth career, for James VI of Scotland appointed him King's Printer on 9th October 1590 (Edinburgh Bibliographical Society Publications Volume VII pages 122-3 plus other articles in other of their volumes).

Waldegrave held this important position as Printer to the Scottish King from 1590 until 1603, a long and continuously active span, save for the year 1601 (possibly through illness).

Aldis lists for 1591 7 extant Waldegrave imprints; for 1592 7; for 1593 9; for 1594 14; for 1595 16; for 1596 8; for 1597 13; for 1598 8; for 1599 11; for 1600 14; for a change for 1601 none for Waldegrave but with 7 by other Scots printers; but then again for 1602 6 by Waldegrave out of a Scots total of 9 for that year. Finally for 1603 Aldis lists a total of 10 Waldegrave imprints out of a total of 26 Scots printings, in that epic year when James VI of Scotland rode slowly (and fearfully) south through York and other English cities to become also James I of England. His mother, Mary Queen of Scots may have been tragically put to death by Elizabeth, yet in the event it was Mary Queen of Scots who posthumously won the battle for the succession to the English throne with the five Stewart regal generations of James, Charles; Charles, James and Anne.

Waldegrave's thirteen years as Scotland's Royal Printer were by no means uneventful nor free from stress. One Sutcliffe in a 1595 book <u>An Answere</u> (quoted by Arber in his <u>Introductory Sketch</u> p 180) states that he had "seen a little pamphlet entitled <u>Martins Interim</u>".

Colson writes:

> Although Throkmorton hoped that his book would be printed in Scotland, it was never set in type. It is understandable why Waldegrave did not print it, since he was under pressure from James VI to print nothing without license, lest he or the King give offence to Queen Elizabeth and her government. Furthermore, the English ambassador, Robert Bowes, kept a close watch on Waldegrave's activities.

After being royal printer to the Scottish King for six or seven years Robert Waldegrave was nonetheless tried there for treasonable printing, on 2nd February 1596/1597, for making changes or errors in the reproduction of a Scots Act of Parliament. Dickson & Edmond give the text of this case on their pages 400-403. It is in old Scots and is taken from Criminal Trials by Robert Pitcairn Vol.ii p 2. Dickson & Edmond's assessment of the transcript is that at least on this occasion Waldegrave was "entirely innocent of wrongdoing."

In this same treasonable printing acquittal case in February of our year 1597, Robert Waldegrave, although still comfortably under fifty years of age, is referred to as 'waik, in respect of his infirmitie and lang seiknes'.

Nevertheless despite infirmity, Waldegrave was able to return to London in 1603 in safety, following the accession of James I and VI. Although, as mentioned, there are 10 extant 1603 Edinburgh imprints by Waldegrave - or in his name - yet on 11-6-1603 a London Stationers Company entry cites:

> Robert Wal[de]grave Entred for his copie vnder the handes of the Lord Bysshop of London and the wardens The Tenne commaundementes with the Kynges Armes at Large quartered as they are.

However, by 6-8-1604 another Stationers Company document refers to "Robert Wal[de]grave deceased", so that he did not for long continue to enjoy the quiet triumph of his return to the London which had imprisoned him for twenty weeks in 1588, an eventful sixteen years earlier.

It will be recalled that one of the 1588 Marprelate works referred to Robert Waldegrave as having a wife and six small children. While her husband was on the run for the puritan cause through the English countryside and at La Rochelle, she presumably stayed in London with her large, young family. However, it would seem that eventually she too was successful in rejoining her husband after he reached Scotland. Certainly there are two extant 1604 'Widow Waldegrave' Edinburgh imprints Ane act of counsell --- (STC 21959) and a proclamation Search for gold and silver mines (STC 21960).

Leland Carlson in his Marprelate book summarises Robert Waldegrave's Scottish printing career thus:-

> During his thirteen years in Edinburgh, he printed about 120 books, more than the 88 books he printed in England from 1578 through 1589. In Scotland he reprinted books by William Perkins and Sir Philip Sidney, and had the satisfaction of printing Cartwright's long delayed book, *With God in Christ; The Answere to the Preface of the Rhemish Testament*, in 1602, shortly before Cartwright's death. Waldegrave also printed three of the best-known books of James VI: *Daemonologie, in Forme of a Dialogue* (1597); *The True Lawe of Free Monarchies* (1598); *Basilikon Doron* (1599, 1603).

(I have a note that James VI and I also wrote a pamphlet attacking the smoking of tobacco, but this was not seemingly printed in Scotland).

Finally, by way of confession as to this author's lack of impartiality, it will already be clear that Robert Waldegrave is one of my major printing 'heroes' in my self-chosen series of now nine studies into the spread of printing through Britain in the first two centuries or so after Caxton, and I believe many other readers will share my admiration for that 'demned elusive' and efficient, adaptable, practical printer Robert Waldegrave in the heroic Armada period of Elizabethan England.

Indeed Waldegrave now shares a corner of my heart with my other on-the-run printing hero of the next following century, Stephen Bulkley (c.1617-1680) who flitted with his mobile printing apparatus from London to York to Newcastle to Gateshead to York, though in the royalist rather than the puritan cause. I eagerly hope to meet both of them in The Shades (certainly not in Heaven and hopefully not in Hell but most likely in Dante's Purgatory where all go after our mortal apprenticeship here). There will be so many exciting tales of "daring-do" and near-capture to be recounted yet! For example how did Waldegrave succeed in travelling incognito with heavy alphabets of black-letter founders type from Coventry to La Rochelle; and again by what cleverly improvised means did he secretly transport "three or four consignments of books" from France to Throkmorton's house in the English Midlands? - One would dearly love to know chapter and verse.

While in Scotland Robert Waldegrave used this swan device, seemingly that of a male mute swan angrily defending its territory (and so not inappropriate). Round the border are dotted the words: 'GOD --- IS --- MY --- HELPER'. Providence must indeed have frequently smiled on Robert Waldegrave during his exciting and varied series of printing careers.

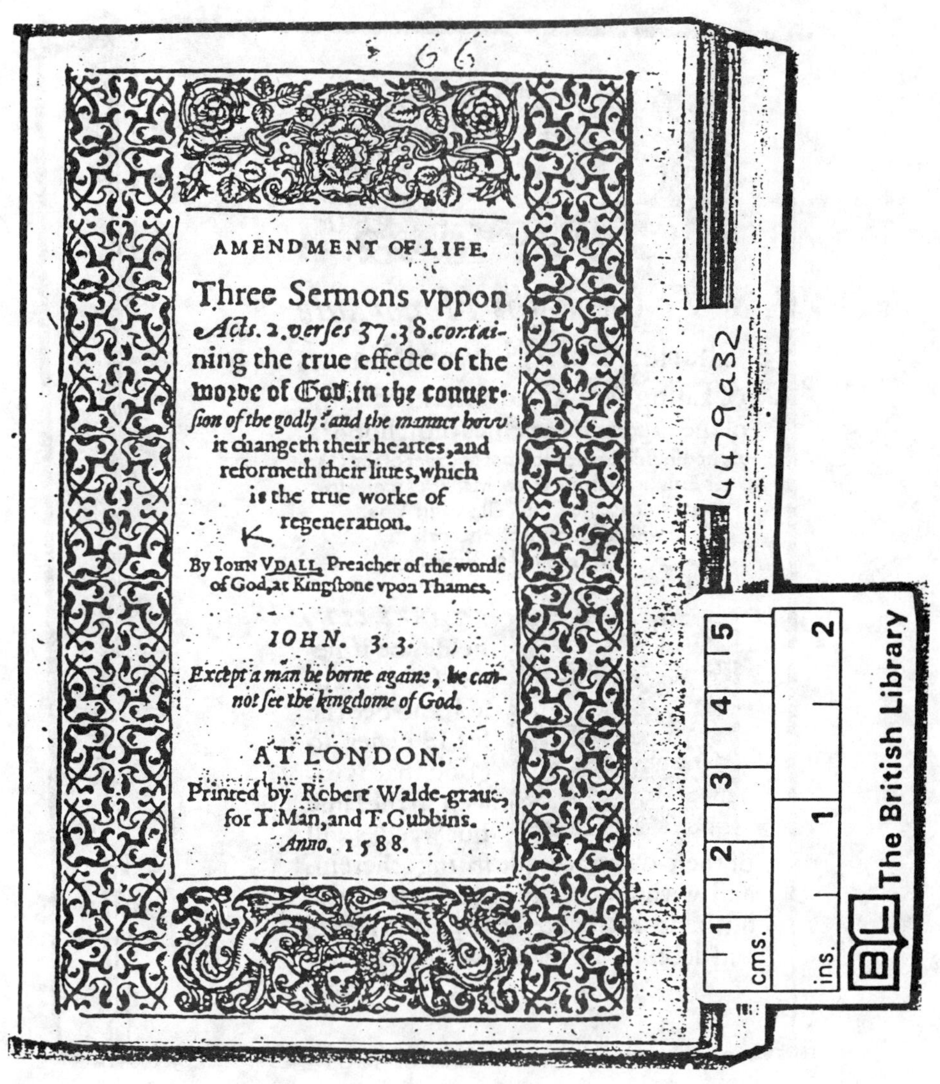

AMENDMENT OF LIFE.

Three Sermons vppon *Acts.* 2. *verses* 37. 38. contai-ning the true effecte of the worde of God, in the conuer-*sion of the godly: and the manner how* it changeth their heartes, and reformeth their liues, which is the true worke of regeneration.

By IOHN VDALL, Preacher of the worde of God, at Kingstone vpon Thames.

IOHN. 3. 3.

Except a man be borne againe, he can-not see the kingdome of God.

AT LONDON.

Printed by Robert Walde-graue, for T. Man, and T. Gubbins. *Anno.* 1588.

Courtesy British Library London
- 4479 a 32 -

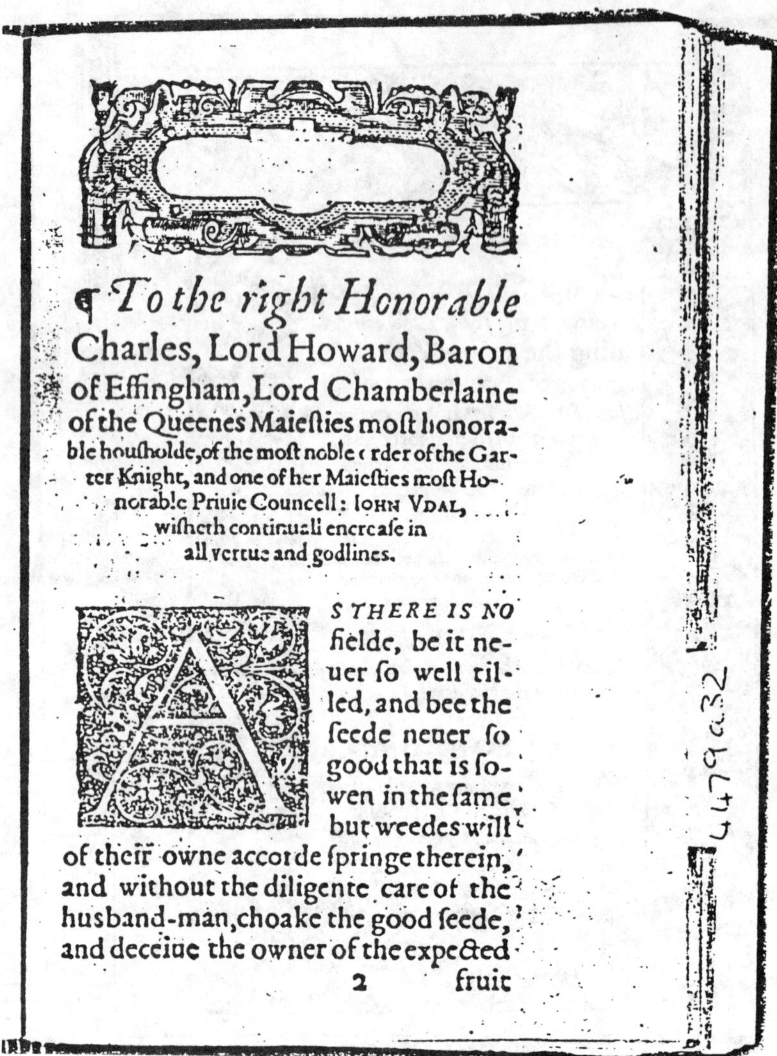

¶ *To the right Honorable* Charles, Lord Howard, Baron of Effingham, Lord Chamberlaine of the Queenes Maiesties most honorable housholde, of the most noble order of the Garter Knight, and one of her Maiesties most Honorable Priuie Councell: IOHN VDAL, wisheth continuall encrease in all vertue and godlines.

AS THERE IS NO fielde, be it neuer so well tilled, and bee the seede neuer so good that is sowen in the same, but weedes will of their owne accorde springe therein, and without the diligente care of the husband-man, choake the good seede, and deceiue the owner of the expected

2 fruit

Courtesy British Library London

- Amendment of Life continued -

THE COMBATE betwixt CHRIST and the Deuill.

Foure Sermons vpon the temptations of Christ in the wildernes by Sathan, *wherein are to be sene the subtile sleightes that the* tempter vseth agaynst the children of God, and the meanes that God hath appointed to resiste him, sanctified to our vse in the example of our Sauiour IESVS CHRIST.

By *Iohn Vdall* Preacher of the word of God, at *Kyngston* vpon Thames.

IAMES. 1. 2. 3. 4.

¶ My Brethren, count it exceeding ioye, when ye fall into diuers temptations: knowing that the triall of your faith, bringeth forth pacience: & let pacience haue her perfect work, that ye may be perfect and entier, lacking nothing.

AT LONDON.
Printed by Robert Walde-graue, for Thomas Man, and William Brome.

TO THE RIGHT honorable HENRY Earle of Huntyngdon Lord Hastings, &c. Of the most Noble Order of the Garter Knight, and Lord President of the Counsell established in the North-partes I. V. VVissheth increase of all spirituall graces, and true honor in this world, and that endlesse glory that lasteth for euer.

WHEN I consider with my selfe (right honorable) that inestimable crowne of immortall glorie which it hath pleased God the father, in his sonne Iesus Christ, to lay 1 vp in store for those, whom (of his mercy 2 and free fauour) he elected vnto him selfe before the foundations of the vvorld: and with all those 3 many tribulations and afflictions, by which his heauenly vvisedome hath appointed them to enter into the same: I see (as in a glasse) before myne eyes, the cause that moued our Sauiour Christ to say, that 4 straite is the way, vvhich leadeth to life and fevv, there by that

(1) Math. 25. 34.
(2) Ephes. 1. 4.5.
(3) Actes 14. 22.
(4) Math. 7. 14.

2 *finde*

Courtesy British Library London
- Combate continued -

EPISTLE DEDICATORY.

[i]f your honor to whō I may dedicate the same, [p]artly for that I might shewe some token of [t]hankefulnesse vnto the same, for your fauour [to]wardes me, and especially for that I am per[s]waded the doctrine, therein conteyned wilbe [a]cceptable vnto you. The Lord of heauen and earth (from whom all true honour commeth) increase the same vnto your duety more and more: for the aduauncement of his glory in this world, and the eternall comfort of your owne soule in the worlde to come. Amen.

Your honors most humble to commaunde in the Lord.

IOHN VDALL.

To the Reader.

THe particular braunches of the whole matter (as you see in this Table) are in number twentie and two, looke what figures you finde any poynte marked with all in this place, and the same shall you finde in the Margent of the booke, where the same matter begynneth to bee handled.

Courtesy British Library London
- Combate continued -

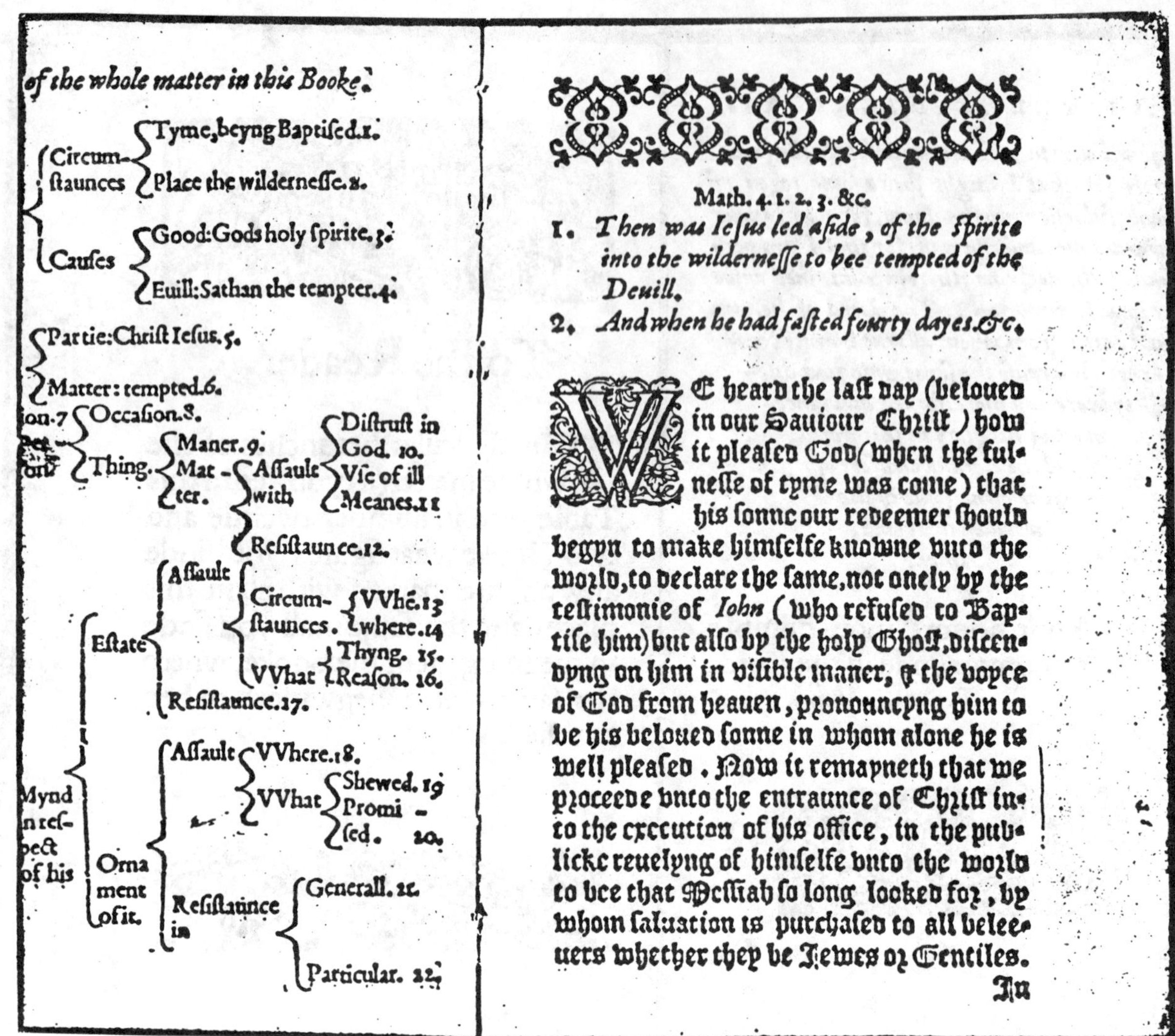

of the whole matter in this Booke.

Circumstaunces
Tyme, beyng Baptised. 1.
Place the wildernesse. 2.
Causes
Good: Gods holy spirite. 3.
Euill: Sathan the tempter. 4.
Partie: Christ Iesus. 5.
Matter: tempted. 6.
...on. 7 Occasion. 8.
Thing.
Maner. 9.
Matter.
Assault with
Distrust in God. 10.
Vse of ill Meanes. 11
Resistaunce. 12.
Estate
Assault
Circumstaunces.
VVhē. 13
where. 14
VVhat
Thyng. 15.
Reason. 16.
Resistaunce. 17.
Mynd in respect of his
Ornament of it.
Assault
VVhere. 18.
VVhat
Shewed. 19
Promised. 20.
Resistaunce in
Generall. 21.
Particular. 22.

Math. 4. 1. 2. 3. &c.

1. *Then was Iesus led aside, of the Spirite*
into the wildernesse to bee tempted of the
Deuill.
2. *And when he had fasted fourty dayes.&c.*

WE heard the last day (beloued in our Sauiour Christ) how it pleased God (when the fulnesse of tyme was come) that his sonne our redeemer should begyn to make himselfe knowne vnto the world, to declare the same, not onely by the testimonie of *Iohn* (who refused to Baptise him) but also by the holy Ghost, discendyng on him in visible maner, & the voyce of God from heauen, pronouncyng him to be his beloued sonne in whom alone he is well pleased. Now it remayneth that we proceede vnto the entraunce of Christ into the execution of his office, in the publicke reuelyng of himselfe vnto the world to bee that Messiah so long looked for, by whom saluation is purchased to all beleeuers whether they be Iewes or Gentiles.

In

Courtesy British Library London
\- Combate continued -

A Prayer.

thy truth may be magnified, thy Sonne Christ Iesus may be aduaunced, and our soules and consciences euerlastingly comforted. Lord we pray thee graunt these graces, not alone to vs that are here present. &c.

Courtesy British Library London
- <u>Combate</u> continued -

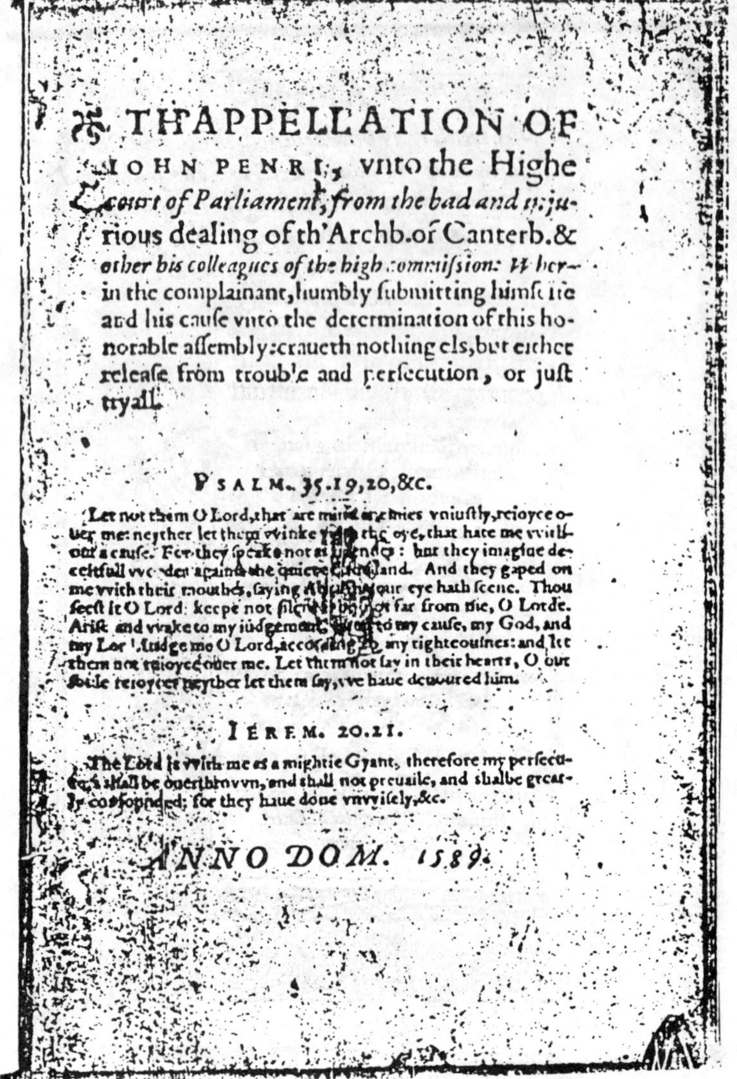

TH'APPELLATION OF IOHN PENRI, vnto the Highe court of Parliament, from the bad and iniurious dealing of th'Archb. of Canterb. & other his colleagues of the high commission: Wherin the complainant, humbly submitting himselfe and his cause vnto the determination of this honorable assembly: craueth nothing els, but either release from trouble and persecution, or just tryall.

PSALM. 35.19, 20, &c.

Let not them O Lord, that are mine enemies vniustly, reioyce ouer me: neyther let them winke [illegible] the eye, that hate me without a cause. For they speake not as [illegible]: but they imagine deceitfull words against the quiet [illegible] land. And they gaped on me with their mouthes, saying [illegible] our eye hath seene. Thou seest it O Lord: keepe not silence [illegible] far from me, O Lorde. Arise and wake to my iudgement, [illegible] to my cause, my God, and my Lord. Iudge me O Lord, according to my righteousnes: and let them not reioyce ouer me. Let them not say in their hearts, O our soule reioyce: neyther let them say, we haue deuoured him.

IEREM. 20.11.

The Lord is with me as a mightie Gyant, therefore my persecutors shall be ouerthrowen, and shall not preuaile, and shalbe greatly confounded; for they haue done vnwisely, &c.

ANNO DOM. 1589.

Courtesy British Library

Probably printed in La Rochelle by Waldegrave

Note the long-winded Penry Title & compare with the 'jokey' Marprelate Titles

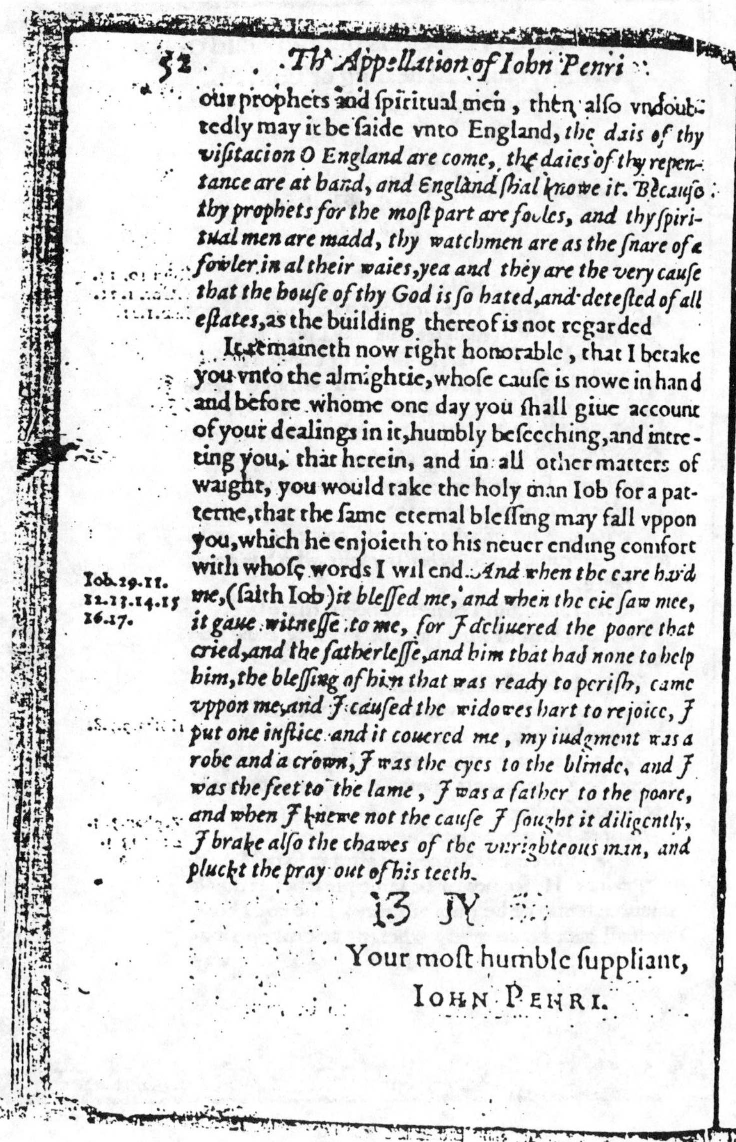

our prophets and ſpiritual men, then alſo vndoubtedly may it be ſaide vnto England, *the dais of thy viſitacion O England are come, the daies of thy repentance are at hand, and England ſhal knowe it. Becauſe thy prophets for the moſt part are fooles, and thy ſpiritual men are madd, thy watchmen are as the ſnare of a fowler in al their waies, yea and they are the very cauſe that the houſe of thy God is ſo hated, and deteſted of all eſtates,* as the building thereof is not regarded

It remaineth now right honorable, that I betake you vnto the almightie, whoſe cauſe is nowe in hand and before whome one day you ſhall giue account of your dealings in it, humbly beſeeching, and intreting you, that herein, and in all other matters of waight, you would take the holy man Iob for a patterne, that the ſame eternal bleſſing may fall vppon you, which he enjoieth to his neuer ending comfort with whoſe words I wil end. *And when the eare hard*
Iob.29.11.12.13.14.15 16.17. me, (ſaith Iob) *it bleſſed me, and when the eie ſaw mee, it gaue witneſſe to me, for I deliuered the poore that cried, and the fatherleſſe, and him that had none to help him, the bleſſing of him that was ready to periſh, came vppon me, and I cauſed the widowes hart to rejoice, I put one iuſtice and it couered me, my iudgment was a robe and a crown, I was the eyes to the blinde, and I was the feet to the lame, I was a father to the poore, and when I knewe not the cauſe I ſought it diligently, I brake alſo the chawes of the vnrighteous man, and pluckt the pray out of his teeth.*

13 JY

Your moſt humble ſuppliant,

IOHN PENRI.

Courtesy British Library London
"Th'Appellation" continued

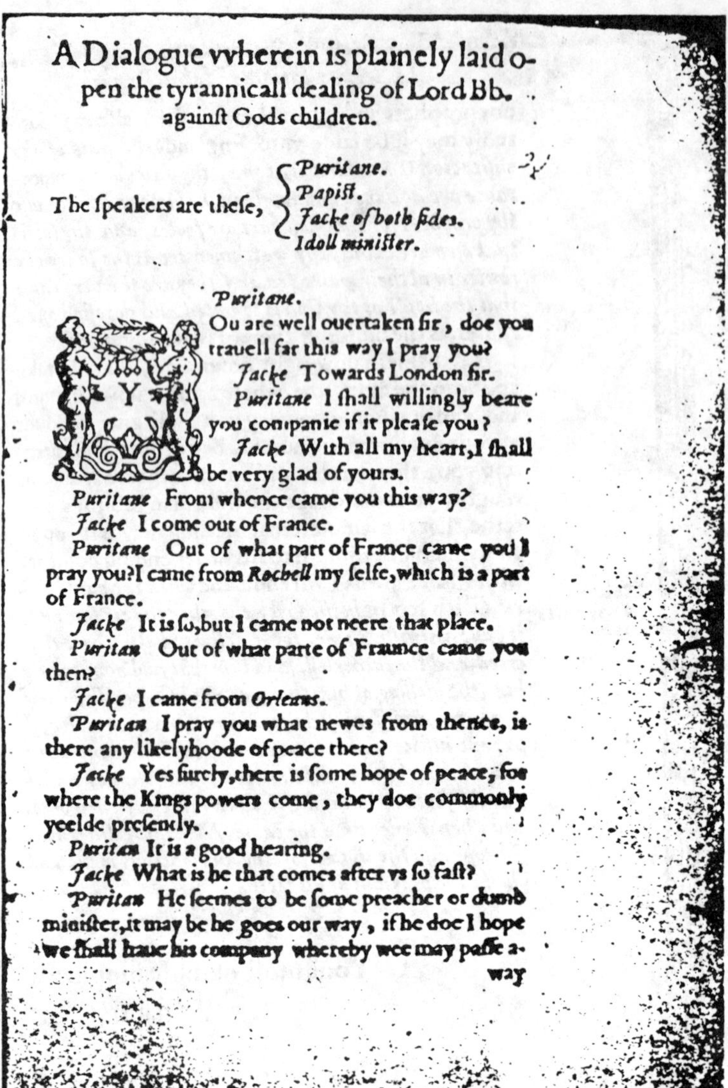

A Dialogue vvherein is plainely laid open the tyrannicall dealing of Lord Bb. against Gods children.

The speakers are these, { *Puritane.* *Papist.* *Iacke of both sides.* *Idoll minister.*

Puritane.

Y Ou are well ouertaken sir, doe you trauell far this way I pray you?

Iacke Towards London sir.

Puritane I shall willingly beare you companie if it please you?

Iacke With all my heart, I shall be very glad of yours.

Puritane From whence came you this way?

Iacke I come out of France.

Puritane Out of what part of France came you I pray you? I came from *Rochell* my selfe, which is a part of France.

Iacke It is so, but I came not neere that place.

Puritan Out of what parte of Fraunce came you then?

Iacke I came from *Orleans*.

Puritan I pray you what newes from thence, is there any likelyhoode of peace there?

Iacke Yes surely, there is some hope of peace, for where the Kings powers come, they doe commonly yeelde presently.

Puritan It is a good hearing.

Iacke What is he that comes after vs so fast?

Puritan He seemes to be some preacher or dumb minister, it may be he goes our way, if he doe I hope we shall haue his company whereby wee may passe a-

way

The beginning (folio A2 *recto*) of *A Dialogue*, an anonymous treatise now proven to be the work of Job Throkmorton, completed in April 1589.

Courtesy Huntington Library California
and Leland H. Carlson

POSSIBLY PRINTED BY R. WALDEGRAVE IN LA ROCHELLE

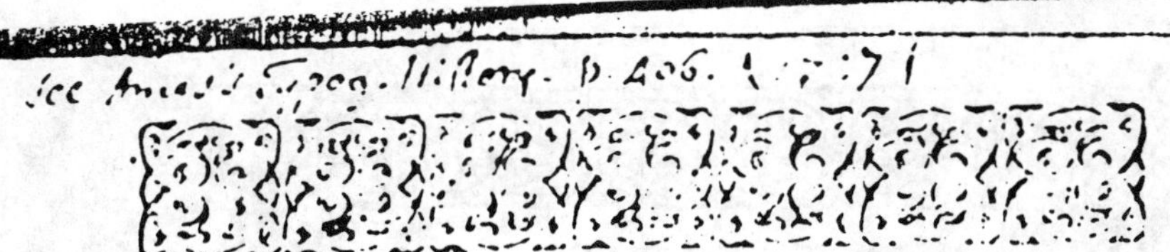

A BRIEFE DISCOVERY OF THE VNTRVTHES AND SLANDERS (AGAINST THE TRVE GOuernement of the Church of Christ) contained in a Sermon, preached the 8. of Februarie 1588. by D. Bancroft, and since that time, set forth in Print, with additions by the said Authour.

THIS SHORT ANSVVER MAY SERVE FOR THE CLEARING OF the truth, vntill a larger confutation of the Sermon be published.

2. PET. 2. 1.2.3.

But there were false Prophetes also among the people, euen as there shall be false Teachers among you, which priuilie shall bring in damnable heresies, euen denying the Lord that hath bought them, & bring vpon themselues swift dãnation. And many shall follow their damnable waies, by whom the way of truth shall be euill spoken of. And through couetousnes shall they with fained words, make marchandise of you, whose judgment long agon, is not farre of.

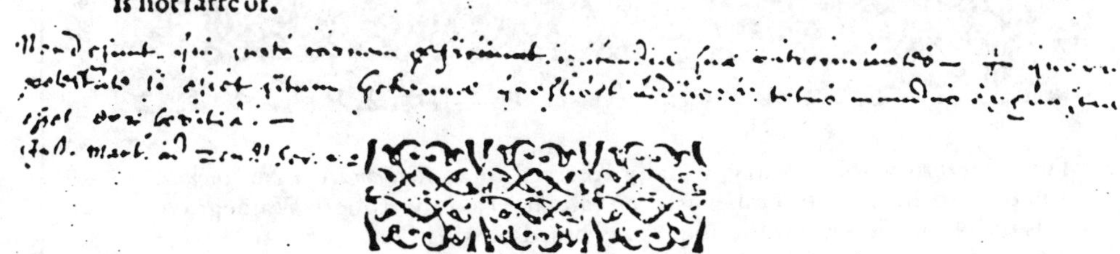

Courtesy British Library London/STC 19603/Aldis 209

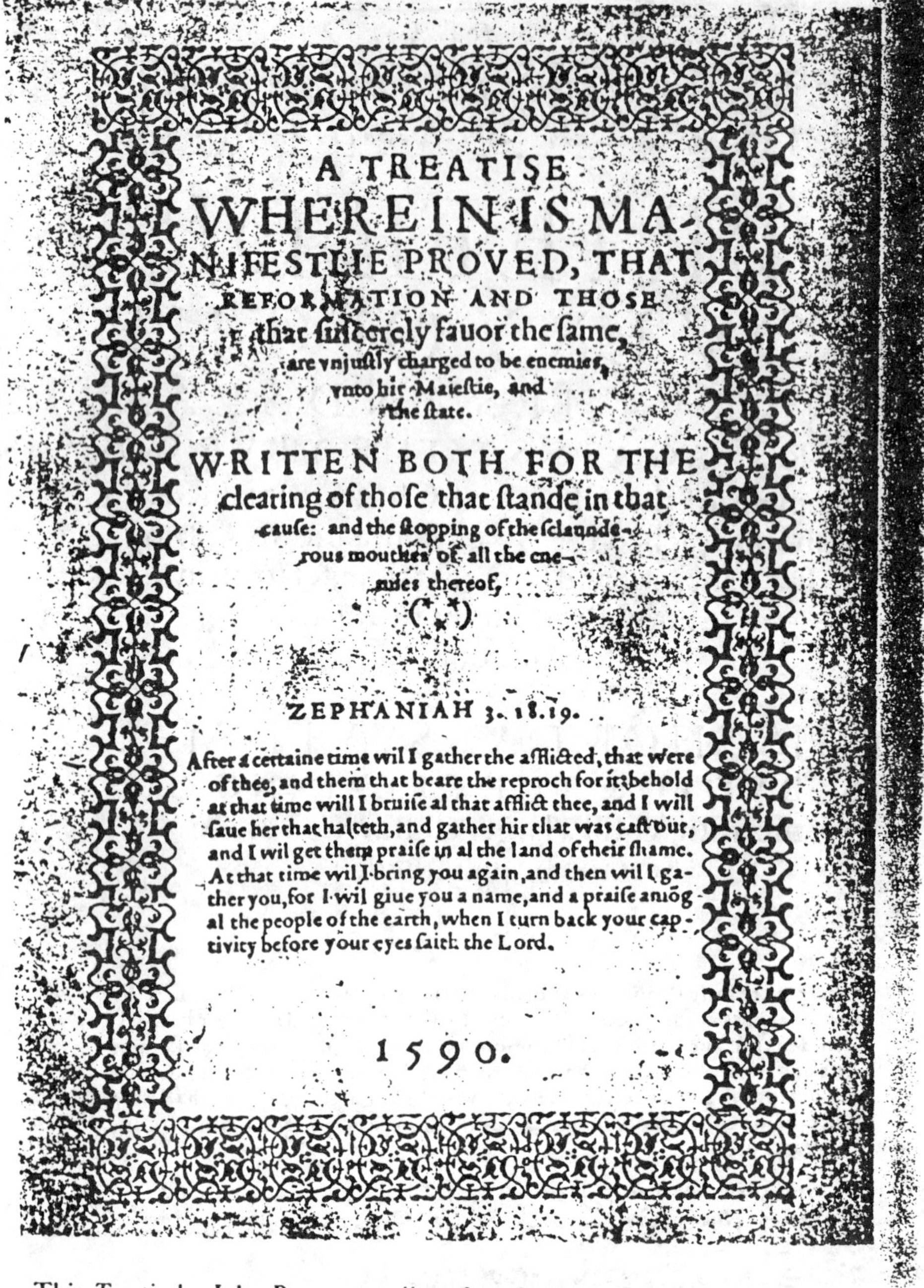

A TREATISE WHEREIN IS MANIFESTLIE PROVED, THAT REFORMATION AND THOSE that ſincerely fauor the ſame, are vnjuſtly charged to be enemies, vnto hir Maieſtie, and the ſtate.

WRITTEN BOTH FOR THE clearing of thoſe that ſtande in that cauſe: and the ſtopping of the ſclaundérous mouthes of all the enemies thereof.

(*,*)

ZEPHANIAH 3. 18.19.

After a certaine time wil I gather the afflicted, that were of thee, and them that beare the reproch for it, behold at that time will I bruiſe al that afflict thee, and I will ſaue her that halteth, and gather hir that was caſt out, and I wil get them praiſe in al the land of their ſhame. At that time wil I bring you again, and then wil I gather you, for I wil giue you a name, and a praiſe amōg al the people of the earth, when I turn back your captivity before your eyes ſaith the Lord.

1590.

This *Treatise* by John Penry, usually referred to as *Reformation No Enemie,* was printed in late December 1589 or January 1590 by Robert Waldegrave at Edinburgh. Penry's indictment was based on this book. This title page is from the copy used by Attorney General Thomas Egerton for Penry's trial before the Court of Queen's Bench. It is at the Huntington Library.

Note again the Long-Winded Penry Title & compare with the 'catchy' titling of Marprelate

Courtesy Huntington Library and Leland H. Carlson

V

PRINTING IN CAMBRIDGE

My Printers Dozen (pages 62-68) included a short summary of the very first printing at Cambridge, in 1520, by John Siberch, otherwise known as Johann Lair von Siegburg. My account was largely based on the 1970 Cambridge University Press publication: John Siberch....... which had been thoroughly researched and written by Otto Trepton and translated by Trevor Jones.

Cambridge's first printed work of February 1520/1521 was an Oratorio or speech of welcome on the occasion of a ceremonial visit to Cambridge by Cardinal Thomas Wolsey. The title-page of Siberch's second work, included Greek type which is held to be the first such usage in a British printed book. Siberch also printed in Cambridge a popular little book on letter writing by the famous continental scholar Erasmus, without the latter's authorisation.

This first Cambridge press was however very short-lived, because, probably by the end of 1523, recently widowered Siberch had returned to the continent with his two daughters of marriagable age; it is thought that in departing he omitted to repay to Cambridge University their initial loan to him of £20.

Eleven years later, 1534 was not only the year of the Act declaring the Royal Supremacy of Henry VIII, but it also occasioned an Act regulating printing which repealed the out-reaching 1484 Act of the one and only Parliament of the maligned Richard III.

As the preamble stated:

> there were but few books, and few printers, within this realm at that time [1484] --- at this day there be within this Realme a greatt number cunning and expert in the said craft of printing, as able to exercise the said craft in all points as any stranger [foreigner].

Thus from 1534, aliens could only sell their wares wholesale to an English-born printer or stationer, and no bound books were to be imported at all.

However in the same year on 20th July Cambridge University obtained Letters Patent from Henry VIII which gave specific licence to the Chancellor, Masters and Scholars to appoint not only stationers but also printers.

For much of this Cambridge printing information I am greatly indebted to Michael H Black's Cambridge University Press: 1584-1984 and to other books shown under 'Cambridge' in my Bibliography. Pages 24 and 25 of M H Black's first-rate book give the text of Cambridge's royal printing charter: there could be three stationers or printers at any one time, who could specifically be either aliens or British-born printers; they could print all manner of books, always provided that they had first been approved by the University Chancellor or his deputy and by three University doctors; and the resulting books could be exhibited for sale wherever they pleased.

When 23 years later, the London Stationers Company was granted its seemingly exclusive printing charter it would appear that there was no thought or realisation as to the mutually conflicting potential of these two printing charters, that of 1534 for Cambridge University and that of 1557 for the London Stationers Guild.

There were to be plenty of hard feelings, verbal conflict and anger between these two august bodies in the decades ahead, but not immediately, because very, very strangely indeed, Cambridge University did not take up its royal printing rights for a full fifty years, i.e. until 1584. In my own reading about these early Cambridge printing years (3 books and 3 booklets) I have not found the least hint of conjecture or data as to a reason for this astonishing half-century printing time lapse. I hope that Cambridge scholarship will redouble its efforts to uncover from local records some new and much needed piece of evidence.

At long last, after some years of ineffective skirmishing and after obtaining the support of Lord Burghley, a Grace was passed by Cambridge University on 3rd May 1583 appointing Thomas Thomas M.A., Fellow of King's as university printer. Incensed, the London Stationers seized Thomas' intended press and types before they could be sent from London, and this triggered a war of words, intrigue and manoeuvring which was to wax and wane far into the following century.

The recommencement of printing in Cambridge in 1584 means that it has had a continuous history of printing and publishing for more than four hundred years to the present day. With the customary modesty of my own university, the Cambridge University Press book points out in its Introduction (p 3) that Oxford University Press did not begin its continuing-to-this-day printing activity until the following year of 1585, so that Cambridge claims to have won that particular printing race "by a length" as it were. The book adds for good measure that Oxford printing did not obtain a charter until 1632 as compared with the Cambridge printing charter of 1534, nearly a century earlier. Oxford's printing longevity stakes will be more than somewhat restored in my subsequent chapter.

Thomas Thomas' successor as Cambridge's University Printer was John Legate I, appointed in the Armada year of 1588. Within two or three years he too was embroiled in argument with the Stationers of London, because he printed at Cambridge a New Testament in (?) 1590, and then a complete Bible in 1591. Since 1577 Bible printing had been held to be the monopoly privilege of the Royal Printer and was jealously guarded.

Legate's 1590 and 1591 Bible printings are considered to be the second and third such to be printed in England outside London. Michael Black writes (p 50):

> Bibles could of course be printed in Scotland, which was still in those days another Kingdom. But until now in England only one book of Scripture had been printed outside London - a New Testament printed at Worcester by John Oswen in 1551, an isolated occurrence which predated Mary's reign, the incorporation of the Stationers' Company and the Royal Printer's Patent given to Barker. Since 1577 it had been accepted that the Bible was not just a London monopoly, but the special privilege of the Royal Printer. Legate's initiative was very much a test case.......

I confirm all the above, save that the Worcester 'isolated occurrence' was plural, as described and illustrated in my The First Printers at Ipswich in 1547-1548 and at Worcester in 1549-1553. At Worcester, John Oswen printed:

(a) The Psalter or Psalmes of David in September 1549 STC 2378 (see my pages 95-96 and 132-134); and (b) Two differing New Testaments (STC^2 2862 and 2862.5) of 1550/1551 known from colophons on incomplete copies. (see my pages 109, 146 and 147).

Falconer Madan in his trilogy Oxford Books gives at page 322 the following useful summary of Cambridge Bible printing (including a 1588 commencement date as will be seen):-

List of Cambridge Bibles or parts of Bibles, to 1750.

1588 24° (Genevan)	1657 8°
1591 8° (Do.)	1657 8° (re-set)
1628 24° (N.T.)	1659 fol.
1629 fol.	1660 fol.
1630 4° (roman)	1661 8°
1630 4° (bl.-letter)	1662 8°
1633 4°	1663 4°
1635 4° (roman)	1664 12°
1635 4° (bl.-letter)	1665 8°
1637 4° (Buck)	1668 4°
1637 4° (Univ. Pr.)	1670 4°
1637 8°	1673 4°
1638 4°	1674 fol.
1638 fol. (standard)	1675 4°
1639 4°	1677 4°
1640 4°	1682 4°
1645 12°	1683 4° (N.T. 1680)
1646 12°	1683 4° (N.T. 1683)
1647 12°	1743 12°
1648 12°	1747 12°

Despite this early Cambridge/London Bible-printing argument, John Legate I, for most of his 32 year long career, succeeded in finding practical working accommodations between the University and the Stationers Company in respect of specific books and classes of books. By 1588 Legate was himself already a freeman of the London Stationers Company. Moreover he married the daughter of Christopher Barker I, the King's Printer, and Legate himself became Master of the Stationers Company in 1604. Although he left Cambridge in 1601, he retained the title of Printer to the University until his death in 1620.

However Cambridge/London printing 'hostilities' resumed in 1620, and in the following year a deputation from the University presented its grievances to King James I as he passed through nearby Royston. In 1623 the Privy Council offered a compromise ruling, not to the University's liking; and although Charles I renewed the University's printing charter on 6th February 1627/8, the economic balance favoured London, for Cambridge needed the full co-operation of the London stationers and book-sellers for the successful selling in good numbers of Cambridge-printed books. By the 1630s Cambridge was printing substantial quantities of school books commissioned by London book-sellers. Cambridge had however through the years eventually won some rights for Bible printing.

In M H Black's words this "heritage of bad blood and frustrated endeavour" between Cambridge University and the London Stationers Company" was to be blown to one side by the national crisis already impending in the [early] 1640s."

> East Anglia was a natural puritan homeland. The Cromwells were local gentry, and Oliver Cromwell, the M.P. for the town, had been at Sidney Sussex [College]. By 1642 the King was asking the Colleges to send him money and plate, while Colonel Cromwell was sending arms to townsmen and training them in their use and himself scouring the countryside to intercept some of the valuables sent to the King in Nottingham........
>
> On 23 August 1642 the House of Commons ordered 'that Roger Daniel, Printer to the University of Cambridge, be forthwith summoned to attend the House, concerning printing the Book set forth in Defence of the Commission of Array; and on 3 September 'ordered, that Mr Daniel..... be injoined by this House, not to print anything concerning the Proceedings of Parliament, without the consent or order of one or both Houses of Parliament: And that he be discharged of further Attendance. In May 1643 Parliament arrested the Vice-Chancellor Dr Richard Holdsworth, and censured him for having licensed the reprinting at the Press of the King's Declarations, first printed at York. (p 69).

As these examples indicate, and as will be seen later, in the graphs of extant title-outputs year by year, Cambridge's civil war printing output continued broadly at the level of the pre-war years, as compared with the sensational surge of civil war title output in the royalist stronghold of Oxford. All the same the Cambridge press continued its fruitful time with Bible printing, including a Greek Bible in 1652 and occasional Polyglot Bibles.

Roger Daniel's appointment as Cambridge University Printer was in 1650 "cancelled for neglect" (a periodic Cambridge printing hazard it appears!). His successor was John Legate II "a freeman of the Stationers Company and to all intents its agent".

This is the same John Legate, who, despite the 1639 Newcastle upon Tyne imprints, was actually in charge of the King's Printers travelling press which dramatically rushed north in response to Charles I's King's Manor York appeal, and who, catching up with the King at The Tyne, was to produce Newcastle's first printing (See my King's Printer at Newcastle upon Tyne in 1639... pages 1-24, and especially that last page).

This John Legate II was appointed Cambridge University printer in 1650, but after leaving very few extant titles, also had his appointment terminated in 1655 "for neglect".

From 1641 the position of the Stationers Company had become legally doubtful

> Star Chamber had been abolished in 1641, and so its supervision lapsed and its ordinances about printing became themselves questionable, though Parliament repeated* some of the old provisions in 1649, notably the restriction of printing to London, York*, and the two [English] universities, though Finsbury was now added [see later]. With no Monarch, there could be no King's Printer, and for a time the Stationers invaded the Bible monopoly, and founded their Bible Stock. But by an order of 1656 Cromwell conferred the Bible monopoly on the partners Henry Hills and John Field. Field styled himself 'one of his Highness's Printers' and 'Printer to the Parliament' so he became the equivalent of King's Printer. What is more, he also became in October 1655 Printer to [Cambridge] University, so he was becoming a monopolist on a grand scale.
> (M. H. Black p 76).

* See our Printing in York p 21.

Field "skilfully" survived the Restoration, and then in 1662 came "the first Licensing Act, which ignored the Stationers' Company and set up a new licensing authority, the Surveyor of the Imprimery and Printing Presses. The Act reduced the national number of master printers. Otherwise it reinforced the provisions of 1637. It continued to recognise the two [English] universities as licensing authorities, but it did not permit them to licence books which were the monopoly of others without the consent of the monopolist. Copies of every book printed were to go to the King's Library and both university libraries".(p 77).

In 1669 John Hayes began his long career as Cambridge University printer, until his death in 1705. He was to pay the University £100 for the privilege of running the press which he owned,helped by "a few journeymen and an apprentice or two". He promised the university authorities not to print heretical nor treasonable matter. He would print his edition - usually 500 or 750 copies of a learned work; 1,500 and upwards for a school book, almanac or Bible" (p 85).

Although he had at the outset promised his university that he would not enter into any undertakings with London printers, yet he needed to sell his book productions, which would not be viable unless he could sell the bulk of his editions to the London booksellers in St Paul's Churchyard, England's major book market. Discreditably, but effectively in practical terms, Hayes did in fact enter into covert arrangements with the London trade so that he kept his Cambridge press busy producing, not only learned works for his own university but also the staple of school books, almanacs and Bibles principally for the dominant London market up to and beyond the end of the 17th century.

CAMBRIDGE UNIVERSITY PRINTERS 1583-1984

1583	Thomas Thomas, M.A.
1588	John Legate
?	John Porter (before 1593)
1606	Cantrell Legge
?	Thomas Brooke, M.A. (before 1608)
1622	Leonard Greene
1625	Thomas Buck, M.A.
	John Buck, M.A.
1630	Francis Buck
1632	Roger Daniel
1650	John Legate the younger
1655	John Field
1669	Matthew Whinn
1669	John Hayes
1680	John Peck, M.A.
1682	Hugh Martin, M.A.
1683	James Jackson, M.D.
1686	Jonathan Pindar
1693	H. Jenkes
1697	Jonathan Pindar
1705	Cornelius Crownfield
1730	William Fenner
1730	Mary Fenner
	Thomas James
	John James
1740	Joseph Bentham
1758	John Baskerville
1766	John Archdeacon
1793	John Burges
1802	John Deighton
1802	Richard Watts
1804	Andrew Wilson
1809	John Smith
1836	John William Parker
1854	George Seeley
1854	Charles John Clay, M.A.
1882	John Clay, M.A.
1886	Charles Felix Clay, M.A.
1916	James Bennet Peace, M.A.
1923	Walter Lewis, M.A.
1945	Brooke Crutchley, M.A.
1974	Euan Phillips, M.A.
1976	Harris Myers, M.A.
1982	Geoffrey Cass, M.A.
1983	Philip Allin, M.A.

Courtesy of
"FOUR HUNDRED YEARS OF UNIVERSITY PRINTING & PUBLISHING IN CAMBRIDGE: 1584-1984"
by David McKitterick.

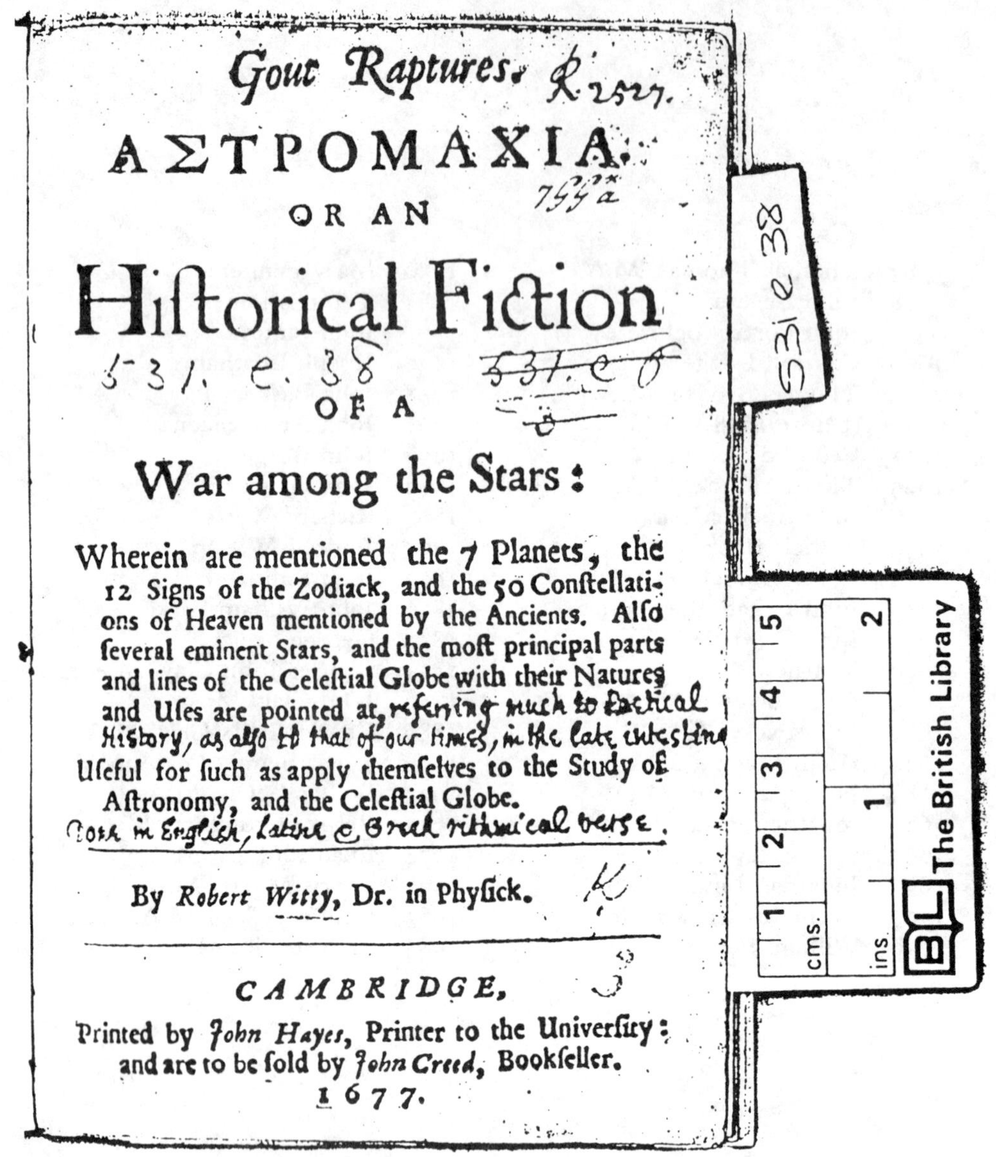

Gout Raptures.

ΑΣΤΡΟΜΑΧΙΑ:

OR AN

Historical Fiction

OF A

War among the Stars:

Wherein are mentioned the 7 Planets, the 12 Signs of the Zodiack, and the 50 Constellations of Heaven mentioned by the Ancients. Also several eminent Stars, and the most principal parts and lines of the Celestial Globe with their Natures and Uses are pointed at, Useful for such as apply themselves to the Study of Astronomy, and the Celestial Globe.

By *Robert Witty*, Dr. in Physick.

CAMBRIDGE,

Printed by *John Hayes*, Printer to the University: and are to be sold by *John Creed*, Bookseller. 1677.

Was this Britain's (or indeed the World's) First Space Fiction?
For another work by Dr Robert Witty, see page 230 of my 'green-back':
Bulkley & Broad: White & Wayt.

Courtesy British Library London

VI

PRINTING AT OXFORD

For this description I am greatly indebted to the late Harry Carter for his History of the Oxford University Press to 1780 (1975); to Nicolas Barker for his Illustrated History, The Oxford University Press and the Spread of Learning: 1478-1978 (1978); and to Paul Morgan editor of the Exhibition Catalogue: Printing and Publishing at Oxford: the Growth of a Learned Press: 1478-1978 (1978).

In my Printers Dozen, I showed on page one the title-page of Oxford's first printed book. This is clearly dated 1468 in latin ("M cccc lxviii") which if true would ante-date William Caxton's 1476 arrival in Westminster from Bruges. The scholarly consensus is that this was a 'thumping' error by Oxford's first printer Theodoric Rood,on the title-page of his first Oxford-printed book, and that an additional 'x' should have been inserted making the date 1478. A main reason for believing this is that Rood's next following books are dated 1479, 1480 and 1481. And yet **is it not** strange that this allegedly inaccurate date was not manuscript-corrected in any of the thirteen surviving copies? (Barker p 4).

Carter lists on pages 10-11 eleven partially or wholly extant works produced at Oxford by Rood between the years 1478 to 1486/7. Then early in the following century another printer John Scolar worked at Oxford for less than a year in 1517 and 1518, with his even more temporary successor Charles Kyrforth, producing one extant book in 1519 or 1520. Again Carter lists the 8 extant titles on his page 13.

There was then printing silence at Oxford for 65 years, until 1585.

It is frequently stated, but incorrectly, that the 1557 London Stationers Company Charter permitted there also to be one press each at Cambridge and at Oxford. Not so. In the preceding Cambridge chapter we have noted that prior to that date, Cambridge University had obtained in 1534 its own printing charter from Henry VIII. Very strangely Cambridge did not stir itself to take up these printing rights for an unbelievable fifty years, until 1584. Nevertheless when they belatedly did so, they were able to resist the instant challenge of the London Stationers, by quoting their 1534 Letters Patent wide-ranging wording.

Equally strangely, Oxford University did not use the obvious 1534 opportunity of their rival or brotherly university obtaining a printing charter. Thus when Oxford University produced its next book in 1585, its printer and so its university authorities were in an even more unprotected position vis-a-vis the Stationers Company's jealously guarded monopoly than in the case of Cambridge.

It was in fact in 1584, the very year of Cambridge's printing recommencement, that Oxford University lent £100 to Joseph Barnes, a local bookseller, to enable him to set up his own printing press. Four months after this loan the University appointed a committee to investigate printing, and through the Chancellor, Robert Dudley, Earl of Leicester, obtained a Star-Chamber decree in 1586 empowering the University to allow one press and one apprentice in Oxford.

For these and the following facsimile pages I am indebted to Paul Morgan's excellently clear and concise Introduction to the above-mentioned 1978 "Catalogue of an Exhibition".

Barnes lived in a house on the north side of High Street, at the west end of St. Mary's church, and worked there as University printer until his retirement in 1617, issuing some 300 well-printed books for which he had a good stock of type acquired from French typefounders working in London. His type-stock included Greek and some Hebrew, and his output, though mainly theological, included such notable secular publications as the first Bodleian Catalogue, 1605 and Captain John Smith's *Map of Virginia*, 1612.

Barnes regularly used the University arms on the title-pages of his books, and described himself as Printer to the University, but his activities were in fact independent: the business was his own and he alone bore all the hazards of financial profit and loss. In this independent, and to some extent uncontrolled position, Barnes was succeeded by two printers who shared a press, William Wrench and John Lichfield, the latter the first representative of a family whose members were styled Printer to the University continuously, working with very varying efficiency, up to 1749. Barnes's successors did less credit to the University and were more obviously uncontrolled, a state of affairs which became critical in 1627 when the two printers Lichfield and Turner broke up their partnership and set up two independent presses, a clear contravention of the intention of the Star-Chamber decree of 1586. This was the immediate occasion of Archbishop Laud's intervention, as Chancellor of the University, to regularize the position by obtaining in 1632 Letters Patent which gave the University a right to appoint three printers, each (by amplification in 1633) allowed to have two presses and two apprentices. But this was the merely mechanical aspect of Laud's great ideal of a university press which should print Greek manuscripts, notably those in the Barocci Collection which he had himself inspired the third Earl of Pembroke to give to the Bodleian in 1628. The next extension of Laud's great plan was the Royal Charter which he obtained in 1636, giving the University the right to print all manner of books. This grant included the printing of Bibles, almanacs, and grammars, formerly the monopoly of the King's Printers and the London Stationers' Company; but the University made no immediate attempt to enter into competition with the Stationers' Company and surrendered its newly acquired right to the Company in return for an annual payment of £200.

This agreement not to print Bibles and grammars was the first of a long series of 'covenants of forbearance' which lasted, with some interruptions, up to 1780. The immediate effect was that the income of £200 made it possible for the University to buy matrices for Hebrew and Arabic type in Leyden in 1637. Laud was delighted at this progress towards the kind of press he had in mind, but the increase in type available was not always matched by an increased efficiency or discipline in the printers. Lichfield printed the *Epistola ad Corinthios* of Pope Clement I, 1633, with the Greek type given to the University by Sir Henry Savile in 1619, 'a work of genuinely modern scholarship', but the other considerable Greek project of these years, the edition of the chronicle of John Malalas, was completely frustrated by the 'peevishness or extreme sottishness' of the printer William Turner. Between the retirement of Joseph Barnes in 1617 and the outbreak of civil war in 1642 there were indeed some notable publications issued in Oxford: several editions of Robert Burton's *Anatomy of melancholy*, the Bodleian catalogue of 1620, and the first English translation of the Latin version of Bacon's *Advancement of learning*, 1640; but no considerable new work of scholarship was issued until, after the outbreak of war, James Ussher, Archbishop of Armagh and Gerard Langbaine, Provost of Queen's College, put learned works to press and ensured a fuller use of the University's stock of Greek, Hebrew, and Arabic type. Their achievement was built on Laud's foundations, and the stock of type was further increased by purchase of more Hebrew and Arabic in 1655–7 and a large fount of Anglo-Saxon in 1656. Then the University printers were equipped to produce that remarkable series of publications in oriental languages and mathematics which distinguished Oxford during the Commonwealth, and the even more remarkable works which embodied the renaissance of Anglo-Saxon studies in Oxford after the Restoration.

In 1658 the University took a step towards direct control of the press by appointing an Architypographus, an academic officer, envisaged in Laud's scheme, who could advance money from University funds to booksellers who would undertake to issue those books which the University wished to see published. For such purposes the University was willing to lend its Greek, Hebrew, and Arabic type, but was still unwilling to run any direct risk of loss or to consider the possibility of profit. For ten years the question of some more formal establishment of a learned press on Laud's model was considered, at intervals and without effect, until in 1668 John Fell, Dean of Christ Church, and Vice-Chancellor of the University, persuaded Archbishop

Sheldon to express a wish that the Theatre, his gift to the University, should serve as a printing-house when not in use for University ceremonies. Now for the first time the University was to have its own premises for a press, and its own equipment. Fell set out his intentions in a letter to Vossius (1671): 'We have it in mind . . . to set up in this place a press freed from mercenary artifices, which will serve not so much to make profits for the booksellers as to further the interests and convenience of scholars.' To that end Fell combined with three distinguished partners who took a lease which relieved the University of all responsibility for the Press and, with their own money and investments from others in the trade, were able to pursue a policy which initiated a heroic age in the history of the Press. Fell did his utmost to provide for all aspects of a learned press: he established a regular type-foundry in Oxford and encouraged the setting up of a paper-mill at Wolvercote; his gifts of Syriac, Coptic, and Samaritan type, added to the Old English, Gothic, and Runic acquired from Francis Junius, and the Hebrew and Arabic bought at Leyden in 1637 made Oxford second only to the Vatican press in range of exotic types, while, with its five presses, it exceeded in potential scale of operation all but three of the London printers. The 1670s saw the production of the third Bodleian catalogue in 1674, Wood's *Historia Universitatis*, 1674, Loggan's *Oxonia illustrata*, 1675, Prideaux's *Marmora Oxoniensia*, 1676, and Plot's *Oxfordshire*, 1677, works which not only brought fame to the Press, but added greatly to the reputation of the University. Fell drew up an even more ambitious programme of publication which was impossible of realization because he commanded neither the finance required nor the means of marketing his products at an economic rate. Years later, Arthur Charlett, Master of University College, recalled this crucial weakness in the Oxford Press: 'The vending of books we never could compasse; the want of vent broke Bp. Fell's body, public spirit, courage, purse and presse.'

After Fell's death his executors made over to the University all his interest in the Press and its equipment, and so at last in 1690 a true University press was established, the University owning the equipment and controlling, or attempting to control, production and distribution through an appointed body of Delegates. These Delegates cautiously decided to 'recommend such Books to the Press as are Vendible as well as Useful', but they still failed to overcome the problems set by slow-moving academic stock. Great books were issued, like Wallis's *Opera mathematica*, 1692–3, Bernard's catalogue of manuscripts, 1697, and the first edition of Clarendon's *History*, 1702, but soon the output of the Press fell off and there was no more urgency in marketing. In 1712, when the Press moved to the newly built Clarendon Building, there were 8,000 volumes in the warehouse, valuable but slow-moving stock, which the Delegates sold off in bulk for a small part of its value, a sad end to the adventurous publishing campaign begun by Fell in 1671.

THE FELL TYPES

A Selection

Double Pica

But all of these Injunctions were the genuine Acts of the *Convocation. The setting forth therefore was not by virtue of*

Great Primer

The Souldiers were in tumult, and seditiously prayed that they *may bee cassiered: not that they soe meant, but by espostulation*

English

And I have therefore spoken so much concerning the Citie in generall, as well to shew you, that the stakes betweene us and them, whose Citie is not *such, are not equall; as also to make knowne by effects, the worth of these men I am to*

Pica

Not that they have the most reason to bee prodigall of their lives, but rather such men, as if they live, may expect a change of fortune, *and whose losses are greatest, if they miscarry in*

Long Primer

Of all the sumptuous edifices which of late years have shot up in Oxford, and adorn'd the habitation of the muses, the new printing-house, commonly call'd Clarendon's Printing-house, strikes me with *particular pleasure and veneration: it is, I do assure your*

Small Pica

SIR. You are now upon a very good Way towards the setting up of a learned Press; & I like your Proposal well to keep your Matrices, *& your Letters You have gotten, safe; & in the*

Brevier

I was loth to write, because I know not how to comfort you, and God knows, I never knew what sorrow ment till now. All that I can say to you is, that you must obey the will and providence of God, and remember, that the Queens *Majesty bare the loss of Prince Henry with a magnanimous heart, and*

Double Pica Greek

ΟΙ μὲν πολλοὶ τῶν ἐνθάδε ἤδη εἰρηκότων ἐπαινοῦσι τὸν προσθέντα τῷ νόμῳ

English Greek

Θράκης πόλεσι, κατέπλευσε μὲν εἰς Ἀμφί-πολιν, ἀλλ᾽ ὅμως μάχης γενομένης Βρασίδας

Long Primer Greek

Περὶ τῶν πέντε ἀστέρων τῶν καλουμένων πλανητῶν, διὰ τὸ κίνησιν ἔχειν ἰδίαν αὐτοῖς, λέγονται δὲ θεῶν εἶναι πέντε· πρῶτον μὲν Διὸς φαίνοντα μέγαν, ὁ δεύτερος ἐκλήθη μὲν

Pica Black Letter

Notwithstanding what hath been said, let none that are godly take occasion hereby to entertaine too high conceipts of themselves

Courtesy: The Oxford University Press and the Spread of Learning by Nicolas Barker.

The Civil War brought great changes to the Oxford Presses, for from the autumn of 1642, King Charles I made Oxford his main headquarters and relied on them for his printing needs.

In January of our 1642 Charles I had left London for the last time as a free man and took a travelling press with him as he journeyed north, having proved its practical usefulness to him on his northward journey of 1639. The King entered York on 18th March of our 1642 and was based there for five months until on 16th August he left for Nottingham where he raised his standard and declared war on the parliamentarians who would not recognise his divine right to rule. During this time the travelling press was busily employed in St William's College at York, producing 74 imprints dating from 26th March 1642 to 7th September of that same year, as shown in the travelling printers contemporary catalogue which I reproduced, courtesy of the National Library of Scotland on pages 32 to 37 of my <u>The King's Printer at York in 1642 and at Shrewsbury in 1642-1643</u>.

Charles I with his army moved westward from Nottingham to Shrewsbury which he reached on 20th September, and then after less than three weeks, on 12th October, moved southwards for the first and undecisive battle on 23rd October at Edgehill near Banbury.

For unexplained reasons the King's travelling printers were delayed in reaching Shrewsbury, in fact until after their Monarch had departed southwards. The same contemporary catalogue records their first Shrewsbury printing as 17th October 1642, and printing continued there in the seeming safety of the walled city of Shrewsbury with its encircling loop of the River Severn until their probable final Shrewsbury imprint dated 16th August 1643 (see Shrewsbury section of the contemporary catalogue reproduced on my pages 38 to 44). The number of known Shrewsbury titles printed by this royal press is 96 of which I have traced only 20 as being extant in the 1980s. This press eventually removed in the late summer or early autumn of 1643 to Bristol, following the Royalists' capture of that city on 26th July 1643.

There the press continued printing for the Royalist cause until Fairfax and Cromwell led the storming of Bristol which surrendered to Parliament on 11th September 1645. Yet, in this sometimes civilized Civil War, the travelling royalist press was allowed to depart in working condition and journeyed unscathed to Exeter, where Royalist printing continued until the final surrender of Exeter, to Parliament on 29th March 1646. That was the month of the Royalist collapse of its final stronghold, the South West of England; for earlier in March the future Charles II had left Land's End by ship for the Scilly Isles, Jersey and the continent of Europe; and also in March the royalist commander Sir Ralph Hopton had finally surrendered to Parliament at Truro.

All of this stirring political and printing activity is described in detail and fully illustrated with virtually all extant imprint facsimiles in my two King's Printer volumes (see back cover).

Why did the travelling royal press stay so long in Shrewsbury, for upwards of a year? The answer, though speculative, seems clear. After the inconclusive October 1642 battle near Banbury, King and Army withdrew to the royalist stronghold of Oxford, where there was already an active printer, Leonard Lichfield, at the ready to carry out the King's printing requirements.

Oxford ('Home of Lost Causes') was the strongly defended headquarters of King Charles I and his Army from autumn 1642 until its final surrender on 24th June 1646, only three months after the royalist collapse in the South-West. However, over a month before the Oxford surrender, Charles I had made a risky, secret horseback ride, with just a few companions northward from Oxford, to Newark on the River Trent where he surrendered his royal person to the occupying Scots Army. After all, like his Father (James VI and I) Charles was King both of England and of Scotland. His immediate removal by the Scots northward to Newcastle upon Tyne, where shortly the King had close personal connections with another ardent royalist printer, Stephen Bulkley, is fully described, with imprint facsimiles in my Bulkley and Broad, White and Wayt volume.

So, while Oxford was the Royalist Civil War headquarters from autumn 1642 until June 1646, what printing service was Leonard Lichfield (with his authorised three printers, two presses and two apprentices) called upon to make to further the King's cause, which was every bit as much a civil war of ideas and ideals, as of sword and cannon.

In Nicolas Barker's words (p 13):-

> The removal of King and court to Oxford doubled and redoubled the amount of work that came to them, for a pamphlet printed in the Royalist interest in Oxford would appear twice, once with an Oxford imprint and again with a false [London] imprint to show the King still had friends there; Parliament, not to be outdone, would alter the text in its own interest and issue it again with two imprints; each might in turn produce a formal rejoinder from the other side. Altogether, over 1,000 pieces came from the Oxford Press between 1641 and 1650, but silent enim inter arma leges - few of them were works of scholarship in the sense that [Archbishop and Chancellor] Laud intended.

"The period of Oxford printing after the Royalists' surrender of the city in 1646 until the Restoration [1660] is remarkable for publications in the 'oriental' languages and in mathematics" (Carter p 37).

An indication of Oxford's extant title outputs year by year, is given in graphical format later. Compared with a 1641 listing from two sources of 44 titles, this leaps for 1642 to 293 titles and for 1643 to 313 titles, reducing for 1644 to 186 titles and further dropping for 1645 to 97 titles. Comparative totals for 1646 were 65 titles: 1647 57 titles; 1648 40 titles and 1649 18 titles, the year of the King's execution. Whilst these individual Oxford title-outputs will not be the correct title totals of the revised Wing listings, yet they serve to give an indication as to the frenetic activity in the royalist cause from 1642 by Leonard Lichfield's presumed two presses at Oxford.

ENGLAND'S FIRST PROVINCIAL NEWSPAPER?

From November 1665, twenty numbers of <u>The Oxford Gazette</u> were printed by Leonard Lichfield, Printer to the University. This was, however, not a provincial newspaper in the usual sense, but was to all intents a government publication: 'Published by Authority', and was started at Oxford because the Court was, for the time, transferred there for fear of the Great Plague of London. Issue number 22 was printed in London and after number 23 the title was altered to <u>The London Gazette</u>, under which title it still continues. (Allnutt p 294).

VII

PRINTING IN IRELAND

In my 'green-back' study A Printers Dozen I described the first book to be printed in Ireland, in 1551, The Boke of the common praier --- after the use of the Churche of England. This was printed in Dublin, in English, by HUMPHREY POWELL. In the previous year on 18th July 1550 the English Privy Council had agreed "a warrant --- to deliver XXli [£20] unto Powell the printer given him by the King's Majestie [Edward VI] towards his setting up in Ireland.

Extant Humphrey Powell Irish imprints are only four, for the scattered years of 1551, 1561, 1564 and 1566/7, (see Facsimile). From these it is nevertheless clear that Powell continued his printing activities in Dublin after Mary became Queen in 1553, and likewise after 1557, for he was himself a member of the London Stationers Company. However, he by no means severed his links with London printing, for in the year of Mary's accession he and a relative William Powell bought a well-known London press from a Protestant printer Whitchurch who had fled to the continent. This enterprise was however costly and the Powells were heavily fined for having failed to raise the promised money.

Next, "Queen Elizabeth had a fount of Irish type made for [printing] the Catechism of 1571". This was in translation "Alphabet of the Irish Language and Catechism --- for anyone who would be obedient to ---- the Queen of this Kingdom, translated from Latin and to Irish by JOHN O'KEARNEY". This controversial work with a faulty Irish 's' (see Printers Dozen) may have been printed, as well as translated, by John O'Kearney, or by a possible relative of the following William Kearney.

WILLIAM KEARNEY, an experienced continental and London printer, was granted a warrant in October 1591 to go to Dublin with his presses with the objective of printing Bibles in the Irish language. He probably did not actually go to Dublin until 1593 when an appeal was made for this major project. Printing of the Gospels began on the premises of recently opened Trinity College Dublin but by 1595 he moved and took employment with the Irish government. (see Dictionaries of Printers..... p 162).

JOHN FRANKE, FRANCKTON (or other variants) printed in Dublin from 1600 to about 1618 and may have learned his undoubted skill in setting the Irish alphabet from prior employment with William Kearney. The 'e' at the end of Franke's name may indicate continental origin. It was John Franke who at long last brought to fulfilment, and to a high standard, the first ever printing in Irish of The New Testament in 1602 (see facsimile) as well as of the Book of Common Prayer also in Irish in 1608.

In 1618 or 1619 Franckton sold his patent rights as Irish state printer to a visiting delegation of the London Stationers Company, and so from 1620 very many imprints appear stating "DUBLIN, SOCIETIE OF STATIONERS" (see facsimile) or more usually "Dublin, Company of Stationers", year after year up to and including 1641, the year before the outbreak of England's Civil War. Between 1618 and 1620 F. Kingston appears on a few Dublin imprints; again in 1635, 1638 and 1640 the surname Crook or Crooke appears along with others.

From 1642 Dublin imprints by WILLIAM BLADEN appear. Dictionaries of Printers ...... Volume II) records that

> In 1618 certain stationers of London formed themselves into a society "to trade in the city of Dublin by vending and selling of books and other commodities ---- to be transported out of England thither, and there to be sold", the books stocked for this purpose being known as the Irish "stock". They appointed William Bladen their factor in Dublin. He was admitted to the Franchise of the city in January 1630/1631 by special grace and on payment of a fine of £10 English money. --- The trading venture of the London stationers had turned out a failure, and the partnership was dissolved in 1639, when William Bladen bought the stock for £2,600. His name first appears in imprints in 1641. In 1647 he filled the office of Lord Mayor of Dublin, and during the Commonwealth he acted as State Printer. His death took place in Dublin in July 1663.-----

One of my favourite titles of William Bladen is his broadsheet "For the better destroying of wolves ---"issued in December 1654 (Wing I405) and reprinted in 1656 (Wing I406).

In July 1660, John Crooke, was appointed Printer General in Ireland with a fee of £8 per annum and power to print all books and statutes; yet he still retained his London bookselling business, until his death in March 1668/1669.

He was succeeded as King's Printer in Ireland by BENJAMIN TOOKE. Although there is one extant 1671 imprint stating "Typis Regiis et impensis Mariae Crooke" (Mary Crooke being John Crooke's widow) Benjamin Tooke's imprints run steadily year by year, from 1669 until 1678. Benjamin Tooke also was a London Stationer, being admitted a Freeman in February 1665/1666.

Throughout and subsequent to his Dublin printing activity, he also continued his London bookselling business. "He is best remembered as Dean Swift's bookseller, and through Swift's good offices he obtained several offices of profit, including that of Printer to the Queen [Anne] in 1713. (Dictionaries of Printers .... Vol III p 293.)

From 1679 Benjamin Tooke shared the office of King's Printer in Dublin with JOHN CROOKE II, whose father had begun official printing in Ireland in the Restoration year of 1660. These "Tooke & Crooke" joint imprints continued until 1683. John Crooke II died in 1684 intestate, with adminstration granted to his mother Mary, who had continued as bookseller into her widowhood, as seen for instance in Wing I957 of 1670 "Dublin B Tooke, sold by Mary Crooke".

Readers of detective novels may be amused that Benjamin Tooke "took over" from "Crooke I" the Dublin royal printery only to yield it in part to "Crooke II". There is nothing sinister about this, yet it serves as a ready aid to memory. Indeed after the death of John Crooke II, Benjamin Tooke appears solus on Dublin imprints once again.

ANDREW CROOK, another son of John Crook I, in 1686 became Benjamin Tooke's "assign in the office of [Dublin] King's Printer, and after 1689 was sole King's Printer. He worked until 1731 and died [in Dublin] in or about 1732. Initially Andrew had been in bookselling partnership with his mother Mary. He developed into printing in 1685 in partnership with Samuel Helsham, whose first solus imprint was in 1681. Helsham's name ceases on extant imprints after 1689.

Finally for our chosen-date-limit of 1695, the name of EDWARD JONES appears in 6 Dublin imprints all in the year 1690. The only Edward Jones in Dictionaries of Printers......... Volume III (p 174) is that of the London printer of Welsh stock who became King's Printer upon the accession of William III in 1688.

The main purpose of this outline summary is to demonstrate that Dublin printing continued regularly and busily from around 1600 onwards, and somewhat more intermittently in the second half of the previous century.

Additionally there were other Irish printing locations in the seventeenth century, namely at:

CORK : KILKENNY & WATERFORD

CORK:

From Cork, in the deep south of Ireland (near 'Kiss-the-Blarney Stone' Blarney Castle, and at the head of the huge and superb Cobh Harbour) there have come eleven extant imprints, in Clough's listing based on STCI:- 2 for 1648, 4 for 1649 and with individual titles for each of 1650, 1657, 1660, 1675 and 1691. These lack clear printers imprints, save for WILLIAM SMITH in 1660 and JOHN BRENT in 1691. Some of the earlier ones were in support of the Irish nationalist cause.

KILKENNY

Kilkenny lies south and inland from Dublin, and is only about thirty miles north of Waterford (see below). Here between the years of 1646 and 1649 there was very active nationalist printing, "By the General Assembly of the confederate Catholicks", with 44 extant titles in Clough's listing. The only named printers in these imprints are THOMAS BOURKE (1648) and in 1648/9 WILLIAM SMITH (see facsimile) who later moved to Cork, as we have seen. According to <u>Dictionaries of Printers</u>...... Volume II p 29 Thomas Bourke was "a native printer" and the authorised official printer of the Catholic Confederation, who printed first between the years 1643-1649 at Waterford (see below) and then at Kilkenny.

As to William Smith, <u>Dictionaries of Printers</u> .... shows him as printing in Kilkenny and then Cork between our years 1649-1667. "As his name is found on Cork imprints as late as 1690, there may have been more than one printer of this name, perhaps father and son."

WATERFORD

Waterford, in the south of Ireland, with Cork well to the west, lies in the higher navigable reaches leading to the fine estuary of Waterford Harbour, protected by the promontary of Hook Head. Waterford was the scene of Thomas Bourke's first series of imprints: "Printer to the Confederate Catholicks of Ireland" (see facsimile) from the year 1643. Then after Bourke had moved to Kilkenny, there appear four titles printed by PETER DE PIENNE in the years 1551 and 1552.

THE BOYNE

I have a note, as yet unconfirmed, that William III had a travelling printer with him at the Battle of the Boyne near Drogheda in 1690. His name was EDWARD JONES, who brought with him a printing press and type which in 1690 he set up "at the King's Hospital in Oxman-Town."

I hope to be able to study further this interesting subject of Irish printing between the 1640s and the 1690s. Printing in BELFAST began in 1697, with PATRICK NEILL's first stated imprint in 1699.

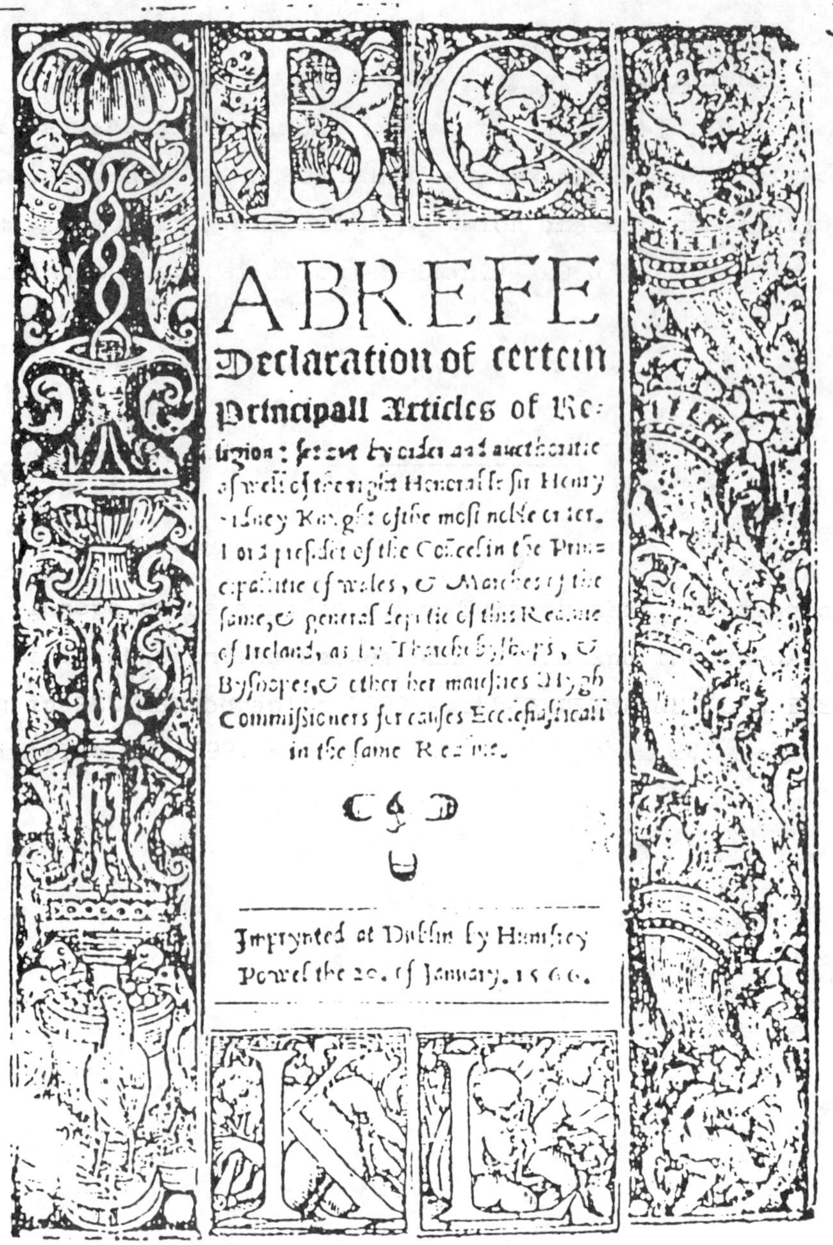

A BREFE
Declaration of certein
Principall Articles of Re-
ligion: set out by order and authoritie
as well of the right Honorable sir Henry
Sidney Knight of the most noble order.
Lord president of the Counsel in the Prin-
cipalitie of Wales, & Marches of the
same, & general deputie of this Realme
of Ireland, as by Tharchebyshops, &
Byshopes, & other her maiesties High
Commissioners for causes Ecclesiasticall
in the same Realme.

Imprynted at Dublin by Humfrey
Powel the 20. of January. 1566.

Courtesy Trinity College Library, Dublin

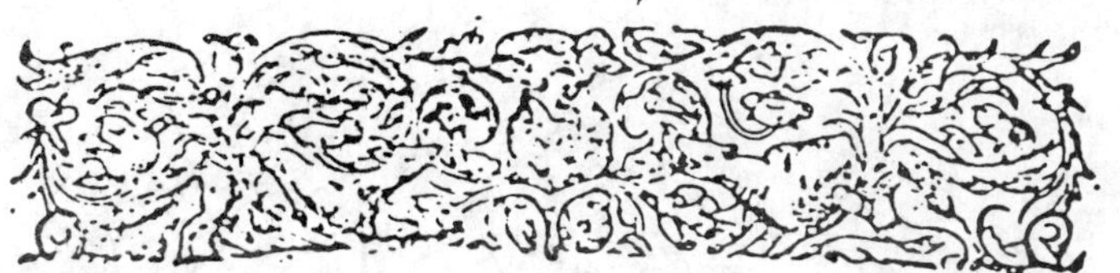

TIOMNA NVADH
AR DTIGHEARNA AGVS AR SLANAIGHTHEORA IOSA CRIOSD.

AR NA TARRUING GU FIRINNEACH ar Greigis gu Gaoidheilg.

RE HUILLIAM O DOMHNUILL.

·Tit. Cap. 2.·

Uers. 11. Do foillsigh grás Dé gu teallruighteach, do beir slánughadh sin do chum na nuile dhaoineadh:

Uers. 12. Agus do beir teagusg dúinne, sá neamh'diadhacht, agus sá ainmianaibh an tsaoghailse do seachna, agus sinn ar mbeatha do chaitheamh dhúinn gu measardha, agus gu coimhthrom, agus gu diadha, sa saoghalsa do láthair.

ATA SO AR NA CHUR A GCLO AMBAILE atha Cliath, a dtigh mhaighistir Uilliam Uiséir Chois an Droichid, ré Seon Francke. 1602.

Title page of the Irish translation of "The New Testament" 1602.

Courtesy <u>Progress in Irish Printing</u> (1936)

AN
Act for the graunt
of one entier Subsidie by
the Temporaltie.

DVBLIN:
Imprinted by Iohn Franckton Printer
to the Kings most excellent Maiesty
Anno 1615.

Courtesy Trinity College Library, Dublin

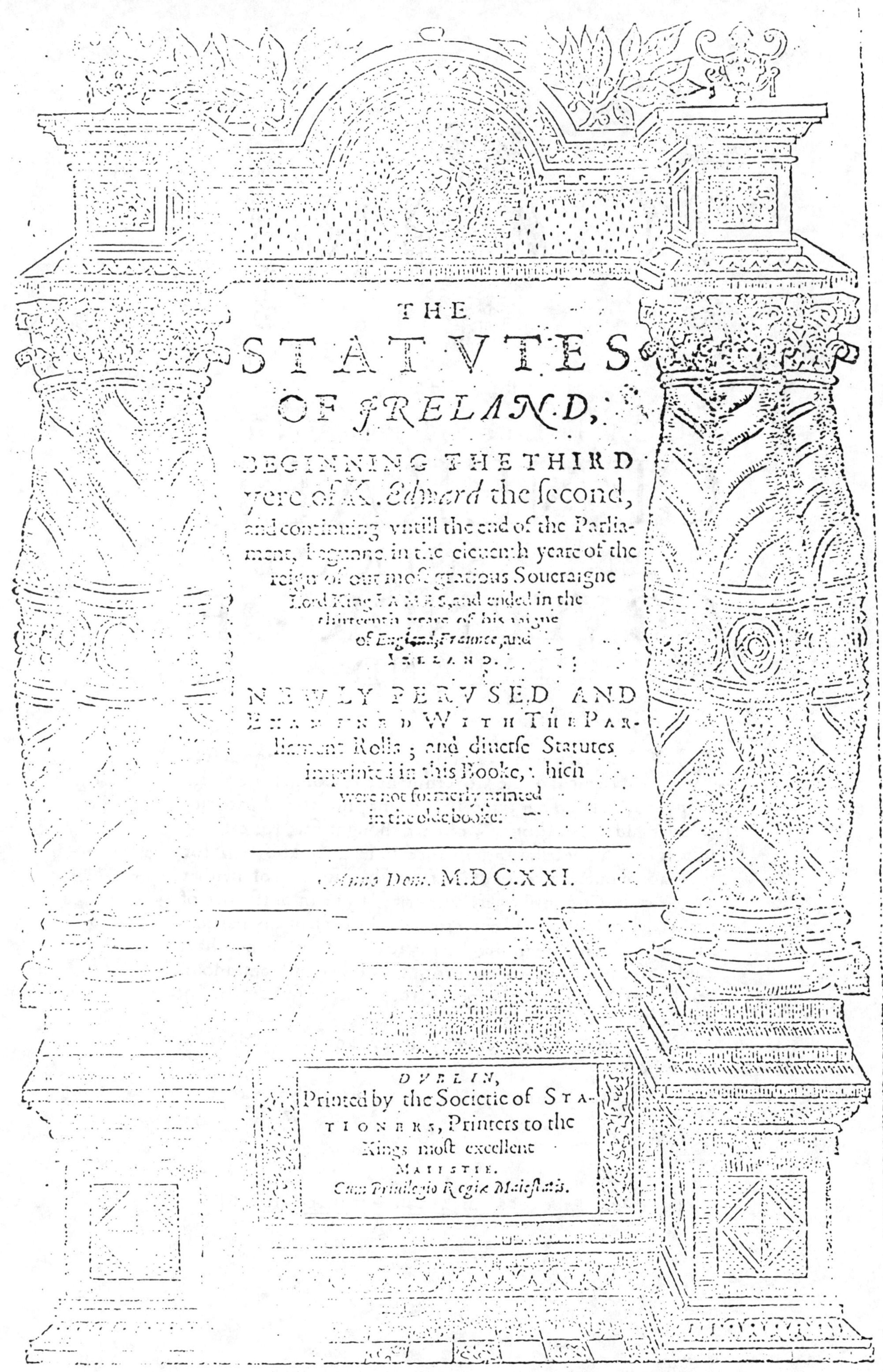

THE

STATVTES

OF IRELAND,

BEGINNING THE THIRD
yere of K. Edward the second,
and continuing vntill the end of the Parliament, begunne in the eleuenth yeare of the reign of our most gratious Soueraigne Lord King IAMES, and ended in the thirteenth yeare of his raigne of England, France, and IRELAND.

NEWLY PERVSED AND EXAMINED WITH THE PARliament Rolls; and diuerse Statutes imprinted in this Booke, which were not formerly printed in the olde booke.

Anno Dom. M.DC.XXI.

DVBLIN,
Printed by the Societie of STATIONERS, Printers to the Kings most excellent MAIESTIE.
Cum Priuilegio Regiæ Maiestatis.

Courtesy Trinity College Library, Dublin

BY THE

LORD LIEVTENANT

GENERALL

AND GENERALL

GOVERNOR OF

IRELAND

ORMONDE.

WHereas many waightie affaires concerning the settlement of the Government, & Composure of the Army must take up our tyme, so as we may not attend particular suits and applications, Wee have thought fitt, for easeing suitors from vnnecessary attendance, to let them know that for the space of one & twenty dayes from the date heereof, neither wee, nor the Commissioners authorized by us in pursuance of the Articles of Peace, will enter into the dispatch of any particular bussinesse; not intending heerby to debarr such as may have cause of Complaint for extortions or other misdemeanours tending to the breach of the peace, from petitioning vs vpon that subiect,

Given at Our Castle of Kilkeny the two and twentyeth of Ianuary 1648.

Printed at Kilkenny by *William Smith* in the yeare 1648.

Courtesy <u>Progress in Irish Printing</u> (1936)

A

REMONSTRANCE

OF GRIEVANCES PRESENTED to his most Excellent Majestie, in the behalfe of the Catholicks of IRELAND.

Printed at *Waterford* by *Thomas Bourke*, Printer to the Confederate Catholicks of *Ireland*. Anno Dom. 1643.

Courtesy <u>Progress in Irish Printing</u> (1936)

VIII

PRINTING IN SCOTLAND

In my Printers Dozen on pages 31 to 41 I briefly summarise the early years of Scottish printing, beginning with the text of King James IV's patent of September 1507, requesting "our Lovittis Servitouris Walter Chepman and Andro Myllar Burgessis of our Burgh of Edinburgh" to introduce into Scotland printing and 'Al Stuf Belangand Tharto and Expert men to use the samyne'. Their main work was the Aberdeen Breviary which was indeed expertly printed with words alternating in red and in black.

Also in my Les Deux Pierres: Rouen, Edinburgh, York, after visits to libraries in Rouen and Paris, I listed a series of named Rouen printers who were at work in 1507, and from whose workshops probably came these first-ever printing craftsmen to work in Scotland. 'Les Deux Pierres' signifies two of the prominent Rouen printers of that epoch: Pierre Olivier and Pierre Violette.

After this initial printing effort by ANDROW MYLLAR the practical organiser, and by WALTER CHEPMAN, the financial man, during the short period 1508 to 1510, there was Scots printing silence until 1520.

For much of what follows I am greatly indebted to the two basic standard works on early Scottish printing, namely (a) Annals of Scottish Printing: 1507 to 1610 by Robert Dickson and John Edmond (1890) and (b) A List of Books Printed in Scotland Before 1701 by Harry G Aldis.

There probably was some small printing in Edinburgh by JOHN STORY around the year 1520 (Dickson & Edmond, p 100) but the next sequence of Scots printing recommenced in 1532 in Edinburgh, by an Aberdonian THOMAS DAVIDSON who left but four extant books before his death in 1542, this being also the year when King James V died.

From the year 1539 there is shadowy evidence (p 152) of a printer in Edinburgh, JOHN SCOT, who may possibly have been the same as a printer in London between the years 1521 and 1537. A Scottish conjecture is that when the English army burned Edinburgh in 1544, John Scot may have fled with his printing apparatus either to Dundee or St Andrews. Ten of John Scot's later works survive, spanning the years 1552 to 1562, with the first St Andrews imprint of 1552 and with others printed some at St Andrews and some at Edinburgh.

In 1562 John Scot was imprisoned by the Edinburgh magistrates, and in 1564 his impounded printing press was delivered to Thomas Bassenden (see later). However Scot seems to have regained its use later for he printed his final 3 extant imprints in 1567 and 1568.

John Scot certainly had possession of types previously used by Thomas Davidson, and another printer, ROBERT LEKPREUIK also used a Thomas Davidson royal arms woodcut (p 199). A Robert Lekpreuik was in 1532 "Banished, by Warrant of the King, furth of the Kingdom of Scotland" but the first sure information about our printer of this name were the beginning of his extant imprints from 1561.

In 1567-8 Lekpreuik was appointed King's Printer "for the space of twenty years" (p 201) with sole rights for certain classes of printing.

Lekpreuik printed busily in Edinburgh up to the end of 1570, with about fifty of his titles surviving; but in 1571 he became involved in disputes between "two factions known as King's and Queen's men". Members of The Queen's party in Edinburgh attempted to arrest Lekpreuik for printing a tract [Perhaps this was STC 22211 "Tressoun of Dumbartane" in the Scottish Public Record Office?].

However Lekpreuik avoided capture of press and person, and made his way to STIRLING, where the court of James VI, the infant King resided. Before 6th August 1571 he produced there Stirling's first piece of printing, and Aldis lists six Lekpreuik Stirling titles for 1571.

In 1572 and 1573 Lekpreuik was busily printing in SAINT ANDREWS, with 15 extant titles. It is thought that he and his mobile press may have left Stirling hastily ahead of the 4-9-1571 attack on Stirling by The Queen's party.

Lekpreuik moved his press back to Edinburgh in 1573 where he worked until January of our year 1574. There he printed a poetical tract without a license; an "Act of 1551 anent printers" was put into force and he was imprisoned in Edinburgh Castle. From then until 1581 he left no extant imprints (a gap of seven years). There were but three in 1581 "after which he disappears altogether from the scene".

> Lekpreuik has no title to be considered a first-rate printer, and none of his works exhibit evidence of the typographic taste and skill which distinguish Davidson's efforts. But during the twenty years in which his press was in operation, he produced [in Scotland] a greater number of individual works than any printer in the sixteenth century, with the exception of Robert Waldegrave (who however was an Englishman) (sic); and to Lekpreuik belongs the merit of having given to the world, for the delight of future generations, so many of those dear old ballads and other rare compositions, every leaf of which is worth more than its weight in gold.

Thus do Dickson & Edmond (p 206) sum up in their own appreciative poetic language the varied career of Robert Lekpreuik. He was Stirling's first printer (1571) and St. Andrews' second printer (1572-1573) - their first having been John Scot from 1552 already mentioned.

Lekpreuik's career illustrates the self-preservation wariness needed by sixteenth and seventeenth century British printers: also the commendable mobility of the wooden hand-press and "al stuf belangand tharto". His two 1571 escapes with his apparatus to Stirling and then to St. Andrews, were in a way the fore-runners of the six or so secret removals of the 1588-1589 Marprelate Press (already described); or of the travelling English royal presses of 1639 and again of 1642 to 1646 with the latter's four printing locations; or again of Stephen Bulkley's two surprise press arrivals in York in 1642 and in Newcastle upon Tyne in 1647. Similarly Lekpreuik's 1571 arrival in Stirling enabled his mobile press to serve the cause of the infant King James VI, who in 1603 was to travel warily south so that Scotland could take over England (said he provocatively!).

The next major figure in Scots printing history was THOMAS BASSANDYNE, a native of Scotland "who acquired the art of printing abroad, having worked both in Paris and Leyden before commencing business in Edinburgh". (Dickson & Edmond p 273).

He is first heard of in 1568 when he was "commanded to call in the books printed by him entitled The Fall of the Roman Kirk, wherein the King was called:-

> 'Supreme head of the primitive Kirk' and to keep the rest unsold till he altered the aforesaid title. He was likewise ordered to delete the "baudie" song "Welcome Fortune" which he had printed at the end of a psalm book, without license......'

Whereas Lekpreuik clearly favoured the cause of the infant James VI, Bassandyne espoused the side of James' equally world famous mother, Mary Queen of Scots. As a result in January 1571/2 he was accused of treasonable practices and denounced as a rebel. However after a time he appears to have offered, along with his friend Alexander Arbuthnot to undertake the printing of a Bible - the first which had been attempted in Scotland - if the Kirk would assist them in the enterprise, which it duly did, as contemporaneously described in detail (see Dickson & Edmond pages 275-278). Earlier Lekpreuik had in 1568 obtained a license to print the whole Bible "in the translation commonly known as the Geneva Bible" but there is no extant result of this.

Eventually, over four years later in 1579, the complete folio Bible was completed (STC 2125). Bassandyne himself did not live to see the completion, as he died in October 1577, yet it would appear from the 1576 dating of its separate title page that he had successfully completed the New Testament.

It would seem that for this massive folio work Bassandyne sought help from continental practical printers, for there is record of a compositor from Magdeburg, named Solomon Kerknett, claiming arrears of his promised weekly wage of forty-nine shillings per week. Bassandyne's Will is quoted by Dickson and Edmond, with details of his printing apparatus including printing ink and "carpentyne" (turpentine) and a very considerable quantity of unsold books!

During 1579, the surviving partner ALEXANDER ARBUTHNOT was appointed King's Printer, just in time for this to be printed with pride on the main Bible title-page. Earlier in that same year he had also been granted a license to print the Psalms, Prayers and Catechisms; it seems clear however that he lacked practical printing knowledge, for both output and standard of work were much lower. Arbuthnot died in September 1585, leaving no books and only a small estate.

The first extant imprint of JOHN ROSS was 1574, and he printed busily from 1575 until his death in July 1580. John Ross' device of the female figure of Truth, holding a lighted candle and a book showing the words 'Verbum Dei', is reproduced here courtesy of Dickson & Edmond (p 328). This woodcut was afterwards used by Robert Waldegrave in 1591 and 1592, and a similar smaller version of it by George Anderson in The Muses Threnodic in 1638.

HENRY CHARTERIS of Edinburgh was a bookseller and a publisher who commissioned works from John Scott, from Robert Lekpreuik and from John Ross; and upon Ross' death it appears that his press was acquired by Charteris. He printed until his death in 1599, after which a younger son ROBERT CHARTERIS struggled to make a printing living. Although appointed King's Printer in Scotland at the end of 1603 (following Robert Waldegrave) Charteris junior was "put to the horn for debt" in August 1609 a year before his death.

THOMAS VAUTROLLIER was a Huguenot (protestant) French printer who was admitted a brother of the London Stationers' Company on 2nd October 1564, but his first extant London book imprint was not until 1570.

"The dilatoriness of Arbuthnot in furnishing the Bibles appears to have annoyed the Assembly" (p 379) so that they urged the invitation of "an active and intelligent printer". Upon Vautrollier's arrival in Edinburgh in 1580 he at first concentrated on book-selling from stocks which he had brought, but he acquired or introduced a printing press and types there in 1584. Yet it appears that Vautrollier only made two extended visits to Scotland, because throughout he continued as a London printer until some time before his death in March 1587/1588. There is evidence that just before Vautrollier's second sojourn in Scotland, he had by his London printing incurred the displeasure of the Star Chamber (p 381). There is also speculation that William Shakespeare visited Vautrollier's London printing workshop in Blackfriars by Ludgate (page 384).

A separate chapter has already been devoted to the career of ROBERT WALDEGRAVE by reason of the many sidedness of his printing activities. Suffice it to state here, that his extant Scots imprints are dated from 1590 (with perhaps one from 1589: STC 17816). From 1590 to 1603 Waldegrave held the position of King's Printer to James VI, with a considerable output of extant works. Yet Robert Charteris also styled himself "Typographus Regius" in 1600 (p 400), this being the year after he succeeded to the business upon the death of his father Henry Charteris; and son Robert, until his death in about 1610, continued to be the Scottish King's printer.

Meantime ROBERT SMITH or SMYTH had also printed in Edinburgh between the years 1592 and 1602, having married as his second wife, the widow of the above-mentioned printer Thomas Bassandyne. Robert Smith was probably the same man who was apprenticed in London for twelve years from February 1564/1565 to Hugh Singleton, the well-known London printer who, in 1567, made a surprise visit to York in order to make a search of the stocks of the York booksellers. (This resulted in four named persons being cited to appear before the Archbishop of York. See our Printing in York pages 14/15 and Robert Davies' Memoir of the York Press, pages 30-33).

Dickson & Edmond's classic work Annals of Scottish Printing covers the period from its commencement in 1507 to the year of 1610. One of the many excellent features in that book is a series of inventories from the Wills of some of these early Scots printers listing book-stocks, household and trade goods. These would repay a detailed examination.

For the bulk of the 17th Century my main source for this continuing bare outline of the sequence of Scottish printers is H. G. Aldis' A List of Books Printed in Scotland before 1700. (1904 & 1970) - not only the 104 pages of Imprint-listing year by year and book by book, but also a concise but very useful "Alphabetical List" of all the Scottish printers (pages 107 to 124).

My main purpose here is to show that in Scotland, throughout this period 1557 to 1695 (chosen for reasons relating to England), there was an overall expansion of Scottish printing and also a continuity of extant Scots imprints from the year 1563 with Aldis' imprint gaps for the 'opening year' of 1557 as well as for 1560, but otherwise continuous: initially in Edinburgh and in later decades also spreading to St Andrews, (1552 and 1572-1573 and 1620-1622), to Stirling (1571), to Aberdeen (1622), to Glasgow (1638), and to Leith (1651).

In the early 17th Century THOMAS FINLASON printed in Edinburgh until his death in 1627. First he purchased Robert Smyth's printing apparatus, being appointed King's Printer for Scotland in 1612. Aldis lists seventy Finlason imprints. ROBERT YOUNG, a London printer, acquired the Edinburgh business from Finlasons' heirs, and thirty-six extant entries of his are dated between 1632 and 1642. Apparently Robert Young stayed in London and sent EVAN TYLER to manage his Edinburgh workshop. By 1641 they were partners and in that year were jointly appointed King's Printers. Evan Tyler's Scottish imprints continued until his departure from Scotland in 1672, there being 365 of his titles listed by Aldis. Their sequence was however interrupted in the 1650s during which time he was declared a rebel by Charles II (before his English Restoration).

In 1651 and 1652 Evan Tyler became the first printer at LEITH, where he was probably succeeded by CHRISTOPHER HIGGINS, who printed Leith tracts between 1652 and 1654, before returning to Edinburgh.

ANDRO HART an Edinburgh bookseller, also commenced printing in 1610 in succession to Robert Charteris, until his death in 1621, after which his press was continued by his second wife and children until 1639, when it was in turn taken over by JAMES BRYSON, previously a bookseller, who expanded into printing until he in turn died in 1642.

ABERDEEN printing commenced in 1622, when an Englishman, EDWARD RABAN, after the one (still extant) 1620 Edinburgh imprint, moved further north in that same year first to St Andrews from 1620 to 1622, and thence northward again to be straightway appointed printer to the Town and the University of Aberdeen. Aldis lists a total of 177 Raban extant imprints. He was printing in Aberdeen from 1622 to 1650, when James Brown succeeded him with the same official printing appointment. Brown occupied the same Aberdeen house as Raban who died in 1658 after eight years of retirement.

At James Brown's death in 1661, JOHN FORBES the elder succeeded to the printing appointment and business. The elder Forbes died in 1675 and the business was carried on by a son JOHN FORBES THE YOUNGER who printed until his death in 1704, after which his widow continued the business. Aldis lists 120 Forbes' family Aberdeen imprints. It is of extra interest that not only was there this long unbroken succession of printing appointment and business from 1622 into the eighteenth century, but also all the families: Raban, Brown and two Forbes rented throughout the same Aberdeen house 'dwelling upon the Market-place, at the Townes Armes'.

JOHN WREITTOUN printed in Edinburgh from 1624 until his death in 1640, with 62 extant entries in the Aldis list.

GLASGOW printing began in 1638 with the arrival of an Edinburgh printer GEORGE ANDERSON, probably upon the invitation of the Glasgow Town Council. In 1636 Anderson in Edinburgh may have printed William Prynne's Newes from Ipswich with Ipswich as its false imprint (see my The First Printers at Ipswich .....) George Anderson "seems to have acquired a considerable part of [Robert] Young's type and ornaments ------; these included several formerly used by Waldegrave and Finlason, among them Ross' two devices " i.e. including the female figure of Truth already illustrated.

"On the death of George Anderson in 1647, the Town Council of Glasgow agreed to continue the subsidy to his widow and children so long as they should carry on printing in the city, but in 1649 they had removed to Edinburgh, and in 1653 were succeeded by Andrew Anderson" (Aldis p 108).

ANDREW ANDERSON, George's son, printed in Edinburgh from 1653 to 1657, but then for the four years 1657 to 1661, he went to print in Glasgow at the invitation of the Town Council "who offered him a yearly pension of 100 marks" (Aldis p 107).

However, for reasons unknown, he returned to Edinburgh in 1661, (being succeeded in Glasgow by Robert Sanders, father and son - see later), and in 1663 was appointed printer to the Town and College of Edinburgh, this being in succession to GIDEON LITHGOW who had died in December 1662. Lithgow had held this post since 1648, following JAMES LINDESAY, who had been appointed to the same office in 1643. It is probable that Andrew Anderson had his printing workshop within the Edinburgh college precincts.

After Andrew Anderson's death in Edinburgh in 1676, his widow "conducted the business in conjunction with, or on behalf of, his son and heir James --- and continued to exercise the offices of King's Printer, and printer to the Town and College" of Edinburgh. For decade after decade, in fact onwards to 1716, there are extant a large series of imprints "Heirs of A. Anderson": "Successors of A. Anderson" with variations on these themes. Aldis lists no less than 960 such imprints and that only extends up to and including his final year of 1700.

Aldis writes (p 107) that "Mrs Anderson came into frequent conflict with the other printers and booksellers in her vigorous and oppressive endeavours to enforce the monopoly conferred by the gift as King's Printer -----". She produced a unique imprint variation in Steele's Proclamations 2508 and Wing Short Title S1497 in 1681, namely "Relict of A. Anderson" and "Printeress to his most Sacred Majesty". I would also greatly like to meet in the Shades this seemingly formidable Widow Anderson, but perhaps not in a competitive printing situation! I believe she may even be globally unique in history in having, even only once, this "Royal Printeress" imprint, (see Facsimile).

As we have seen the title-output from this long-continuing press was substantial, but the extant copies show that the quality-standard was low. Aldis writes: "A considerable portion of [Andrew Anderson's] type and ornaments had been in use in Edinburgh by a succession of previous presses, and are in a much worn condition. His productions, and those of his successors, are among the poorest and most slovenly that proceeded from the Scottish press". (This standard was of course to be rectified for the honour and glory of Scotland by the press of the Foulis brothers in Glasgow, but that was to be in the following eighteenth century).

From 1660 until 1690 there was also what would now be termed a consortium of printers operating as THE SOCIETY OF STATIONERS in Edinburgh. This arrangement began with the Christopher Higgins who had probably printed at Leith for two years up to 1654 (as already described). Between 1655 and 1660 Aldis lists 73 Edinburgh Higgins' imprints. Successively Robert Young, Evan Tyler and Christopher Higgins had come from London, originally to take over the Thomas Finlason business with its royal appointment to print for the King in Scotland. Upon Christopher Higgins' death or retirement in 1660 a varying series of members of the London Stationers' Company seem to have owned shares in an Edinburgh printing workshop with a Scotsman Patrick Ramsay as its overseer. Eventually after this workshop or series of workshops had operated on and off for some thirty years, it was wholly acquired in 1690 by one of the partners, GEORGE MOSMAN, hitherto a bookseller. By the end of that year he had been appointed 'printer to the Church of Scotland and her Assemblies' in spite of Mrs Anderson's opposition, and he continued printing into the following century.

After the 1661 removal to Edinburgh of Andrew Anderson, GLASGOW printing continued with the granting to ROBERT SANDERS I on 23-9-1661 of an annual subsidy of "forty pounds Scots". Aldis (p 119) records 91 extant imprints including 'Printer to the Toun' (1662); 'Printer to the City and University (1672); and 'One of His Majesties Printers' (1683). This last resulted from his having purchased 'George Swintoun's share of the gift as King's printer' probably in that same year.

Sanders died on 12-7-1694 and was succeeded by his son ROBERT SANDERS II. He continued as printer to the City and University of Glasgow, as well as being one of the King's Printers. Aldis (p 120) lists 41 extant titles, but that is only until Aldis' original ending year of 1699, and Robert Sanders II continued printing in Glasgow on into the 18th Century before his death in January 1730.

THE DUTCH PRINTING GROUP AND HOLYROOD HOUSE:

Aldis (p 116) quotes from J Watson's 1713 volume, History of the Art of Printing, that in 1681 Sir Thomas Murray of Glendoick commissioned an Edinburgh bookseller John Cairns to bring across to Edinburgh workmen and materials from Holland to print digests of the Scots Acts of Parliament. Although John Cairns died shortly afterwards, the two Netherlands craftsmen, JOSUAH VAN SOLINGEN and IAN COLMAR continued to print and acquired the Edinburgh business in what would now be termed 'A Management Buy-Out'.

Although the redoubtable Widow Anderson tried to stop their working, an Edinburgh merchant DAVID LINDSAY "obtained a gift from Charles II", and in return Lindsay received a share in what thus became a Scottish-Netherlands printing partnership, with at least one imprint "Lindsay and his partners".

Subsequently Solingen and Colmar bought-out Lindsay's share, with a new, probably Netherlands, printing partner, KNIBLO, who figures in extant imprints of 1683 and 1684. However shortly afterwards this partnership was rescued from debt by being acquired in 1685-6 by an Aberdeen merchant JAMES WATSON, with the Dutch ex-partners seemingly continuing as his employed craftsmen or journeymen.

Aldis writes (p 123) "In February 1686 [James Watson's Edinburgh] premises having been broken into by the populace and his workmen (including THOMAS NOBLE) ill-treated, he was taken under royal protection, and his press set up in the precincts of Holyrood Palace. He was also appointed printer to his Majesties Royal Family and Household, and granted other privileges".

James Watson died in 1687, his printing business being taken over by PETER BRUCE. For 1687 Aldis lists 14 extant 'Holy-Rood House' imprints, with a further 21 such for 1688, but then in December of that year a mob wrecked the royal palace. Aldis (pages 109-110) writes:

> [Peter Bruce was] a foreigner who from 1674 onwards was engaged in engineering and other enterprises (including the manufacture of paper and playing cards) in Scotland. In October 1687 he took over James Watson's press in Holyrood Palace, and in December of the same year was appointed royal printer in succession to him. His printing house was wrecked by the mob which broke into Holyrood on 10 December 1688, and Bruce himself was imprisoned till June 1689. Watson states that [Bruce's] printing materials were sold to the Society of Stationers. His paper mill at Restalrig was acquired by James Hamilton in 1690. The imprints of his books usually run: 'Holy-Rood House, Printed by Mr. P. B. Enginier, Printer to the King's Most Excellent Majesty, for his Houshold, Chappel and Colledge.

The background to this December 1688 looting of Holyrood Palace, was the previous month's 'Glorious Revolution' (to use Protestant parlance) when Dutch William of Orange had landed at Tor Bay between Brixham and Torquay in Devon. Before Christmas of the same year the army of James II had 'melted away', and by then James and his baby son (the future 'Old Pretender') had fled to the Catholic Court of Louis XIV of France.

Many of the active printing personalities appearing in the above outline of the first two centuries of Scottish printing would well repay further study; and this would most definitely include this last-narrated sequence, beginning with the 1681 Netherlands-Scots printing initiative, and ending with this December 1688 destruction of the royal-favoured printing workshop situated actually within the 'holy of holy' precincts of 'Holy-Rood' Palace.

Another subject which would be of interest to study further would be that of printing in the GAELIC language. The first that I have spotted thus far is Aldis' number 3407.3 for 1694, of which there is a unique copy in the National Library of Scotland, namely

Sailm Dhaibhdih : Le Oighreachaibh A. Ainderson: A Ndun-Edin.

By comparison the first Dublin printing in the Irish celtic language was the Protestant Catechism of 1571 (see my Printers Dozen pages 103 and 104). And the first Welsh language production of the Bible was printed in London in the Armada year of 1588 (precisely four hundred years ago) as depicted in the 1st March (St David's Day) special issue of the series of British stamps depicted overleaf:-

18p 26p 31p 34p

FIRST DAY OF ISSUE
1 MAR 1988
YORK

THE WELSH BIBLE
1588-1988

Bishop William Morgan's Welsh translation of the Bible in 1588 is the single most important publication in the culture and religion of Wales. Not only did it help to root Protestantism in Wales, but at a time when the old literary tradition was disintegrating it provided a standard of pure vocabulary and idiom which has safeguarded the quality of literary Welsh over four centuries. Morgan's Bible drew on the earlier work of William Salesbury and Richard Davies who had brought out the Welsh New Testament in 1567, and it was further revised in 1620 by Richard Parry. A new Welsh translation will be published on 1 March 1988.

Date of issue: 1 March 1988

The four stamps were designed by Keith Bowen

Printed in photogravure by Harrison & Sons Limited

Format: vertical
Size: 35mm x 37mm
Perforations: 14 x 14½
Number per sheet: 100
Paper: unwatermarked phosphor coated
Gum: PVA Dextrin

Cover design by Newell and Sorrell

Text by Dr Brynley F Roberts

For information about philatelic services please write to the British Philatelic Bureau FREEPOST Department 85 LHA 20 Brandon Street EDINBURGH EH3 0HN or telephone (031) 556 8661

1988 has been designated 'Year of the Bible' by a number of religious organisations in Scotland. Their symbol is shown below

Courtesy H M Post Office

PRINTERS, BOOKSELLERS, STATIONERS.

EARLIEST DATES OF PRINTING IN SCOTTISH TOWNS.

EDINBURGH	1508	GLASGOW	1638
ST ANDREWS	1552	LEITH	1651
STIRLING	1571	CAMPBELTOWN	1685
ABERDEEN	1622	MAYBOLE	1694

I. TOPOGRAPHICAL LIST.

NOTE.—*The dates of the printers are those of the earliest and latest works noted.*

Aberdeen.

Raban, E.	1622—1649
Melvill, D., bookseller	1622—1633
Brown, J.	1650—1661
Stranghan, D.	1659
Miller, J., bookbinder	1660—1661
V., F.	1661
Forbes, J.	1662—1666
Forbes, J., yr.	1668—1704
Thomson, P., bookbinder	1698—1699

Campbeltown.

No. 2539 (without printer's name)	1685

Dundee.

No. 2393, 'to be sold in Dundee.'	1683

Edinburgh (printers only).

Chepman and Myllar	1508
Chepman, W.	1508—1510
Story, J.	1520?
Davidson, T.	1532?—1542
Scot, J.	1558?—1571
Lekpreuik, R.	1561—1571
(Second press)	1573—1582
Bassandyne, T.	1568—1578
Ross, J.	1574—1580
Arbuthnet, A.	1579—1584
Charteris, H.	1581—1599
Vautrollier, T.	1584—1585
Waldegrave, R.	1590—1603
Smyth, R.	1592—1602
Charteris, R.	1600—1610
Finlason, T.	1604—1628
Finlason's Heirs	1629—1630

Edinburgh—*continued.*

Hart, A.	1610—1621
Hart's Heirs	1622—1639
Hart, J.	1630—1631
Hart, Widow	1631
Raban, E.	1620
Wreittoun, J.	1624—1638
Young, R.	1633—1638
(Second press)	1641
Anderson, G.	1637—1638
Anderson's Heirs	1649—1653
Bryson, J.	1638—1642
Bryson, R.	1640—1645
Bryson's Heirs	1646
Bryson, R. and J.	1641
Young and Tyler	1641—1642
Tyler, E.	1642—1651
(Second press)	1660—1672
Lindesay, J.	1643—1645
Lithgow, G.	1647—1661
Anderson, A.	1653—1657
(Second press)	1661—1676
Anderson and his partners	1671—1675
Anderson's Heir	1676—1694
—— Heirs	1680—1700
—— Successors	1693—1694
—— Heirs and Successors	1694—1716
Higgins, C.	1655—1660
Society of Stationers	1660—1671
(Second period)	1689—1690
H., I.	1663
Swintoun, G., and Glen	1667—1674
Swintoun, G., alone	1669—1678

I reproduce from Aldis pages 105 and 106 listing city by city the Scots 'Printers, Booksellers and Stationers'

Edinburgh—*continued.*

Glen and Trench	1671
Swintoun, G., and T. Brown	1671
Swintoun, G., Glen, and Brown	1671—1676
Brown, T., alone	1674—1678
Glen, J., alone	1675
Swintoun, J., alone	1675—1681
Brown, T., and J. Glen	1678
Brown, T., and J. Swintoun	1679
Cairns, J.	1680
Carron, W.	1680
Reid, J.	1680—1716?
S., J.	1680
Brown, Glen, and Weir	1681
Lindsay, D., alone	1681—1683
Lindsay and his partners	1682—1684
Solingen, J. van, alone	1682
Solingen and Colmar	1682—1685
Lindsay, Kniblo, Solingen & Colmar	1683
Kniblo, Solingen, and Colmar	1684
Colmar, J., alone	1685
Colmar and Gunter	1685
Watson, J.	1686—1687
Bruce, P.	1687—1688
Mosman, G.	1690—1707
Watson, J., yr.	1695—1722

Glasgow.

Sanders, J., bookseller	1625—1642
Wilson, J., bookseller	1634—1635
Anderson, G.	1638—1647
Anderson's Heirs	1648
Neill, J., bookseller	1642—1645
Sandersone, R., bookseller	1654
Anderson, A.	1657—1661
Falconer, J., bookseller	1659—1662

Glasgow—*continued.*

Morison, J., bookseller	1659—1662
Paterson, M., bookseller	1659—1662
Bibliopolarum Typographeum	1661
Sanders, R.	1661—1694
Andrew, J., bookseller	1676
Brown, G., bookseller	1676
Brown, J., bookseller	1676—1685
Dunlop, J., bookseller	1676
Reid, J., bookseller	1676
Scott, J., bookseller	1676
Stewart, R., bookseller	1676
Cunyngham, A.	c. 1680
Hepburn, A.	1689
Sanders, R., yr.	1695—1730
Dickie, W., bookseller	1695—1697
Wilson, J., bookbinder	1699

Leith.

Tyler, E.	1651—1652
[Higgins, C. (?)]	1652—1654

Maybole.

No. 3371 (without printer's name)	1694

Perth.

Lauder, W., bookbinder	1591

St Andrews.

Scot, J.	1552—1558?
Lekpreuik, R.	1572—1573
Raban, E.	1620—1622
Drennane, J., bookseller	1645
Dradoun, G., bookseller	1654

Stirling.

Lekpreuik, R.	1571

THE TAKING OF NEVVCASTLE: OR NEWES FROM THE ARMIE.

After the using of all faire meanes, for reducing the Towne of *New-Castle* unto the obedience of King and Parliament, and their obstinate refusall, of such conditions, as better could not have been expected by people in their case: Yesterday, being Satterday, the ninteenth of *October*, our Batteries began to play by the breake of day: And toward three a clocke in the afternoon, foure Breaches were made in the Wall, Our Mines, one at *Close-gate*, and three at *Sand-gate* were sprung, and served exceeding well. Then did wee make an universall assault: The Breaches by the Mines gave the easiest entrie: The Breaches by the Canon abode longer dispute, being of harder accesse: Before five a clocke all the Breaches were entred. The Major, Ministers, and our Countrey-men reteired to the Castle, where they hope to make their quarter, but it is not likely they can hold out long.

In all the hote service (so farre as we know) we have not lost an hundreth men, some whereof are Officers, *viz.* Lievetenant Collonell *Hume*; and his Major *Hepburne*, and Lievetenant Collonell *Hend[illegible]son* a Reformeir.

Our people were so mercifull, though they had received some losse, that they killed very few, after they were entred: As for the other medlings of the Souldiers, what it was, wee know not; but sure they have laid their hands about them. *Lodevicke Lindesay*, some-time designed Earle of *Crawfurd*, and others, are entred into the Castle also, and the Lord *Rayes* taken by Colloney *Ray*. The Castle sounded a Parlie, but it was not accepted by our *Generall*.

from New-castle *the twentie of October.* 1644.

Printed at *Edinburgh* by *James Lindesay*, 1644.

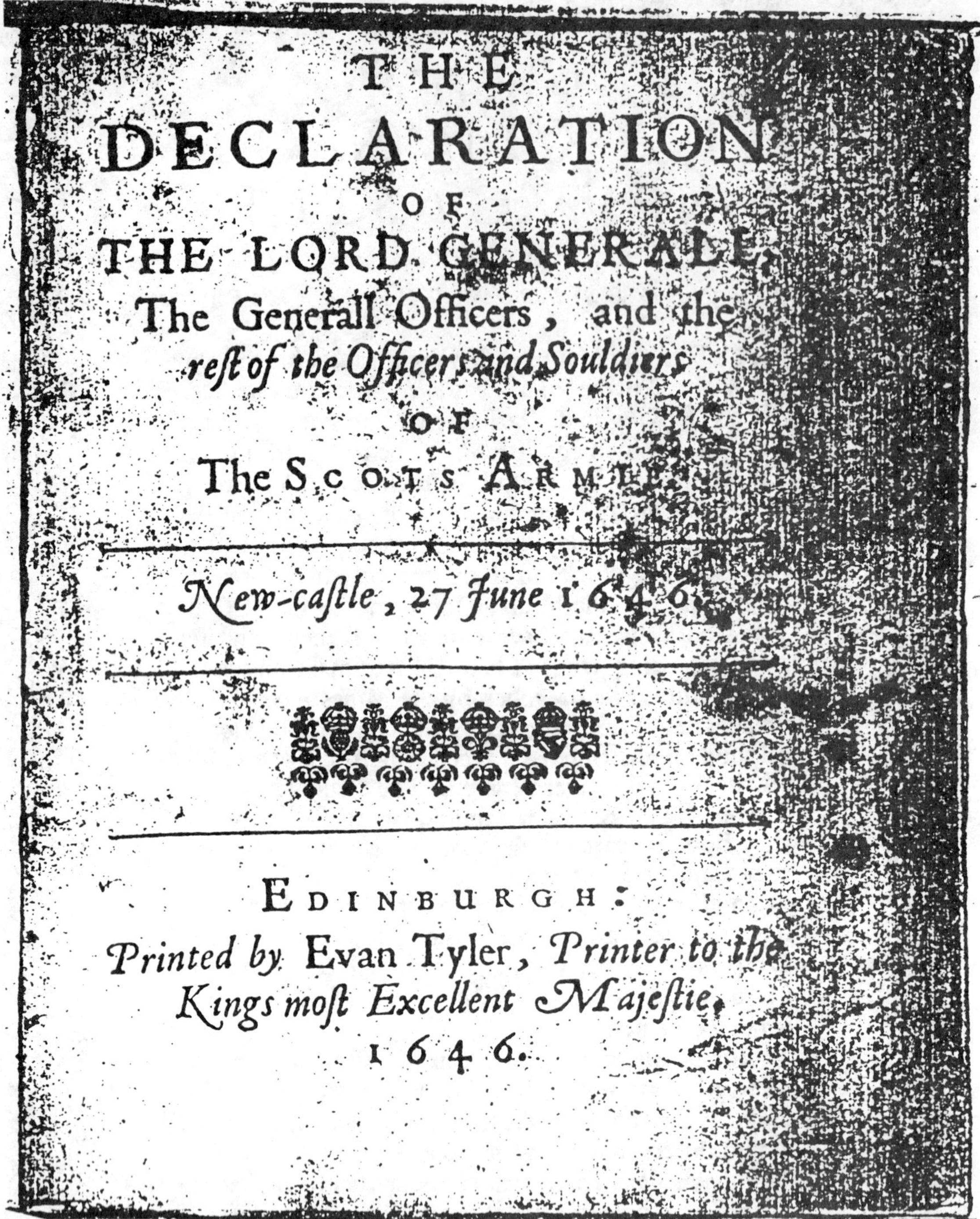

THE
DECLARATION
OF
THE LORD GENERALL,
The Generall Officers, and the
rest of the Officers and Souldiers
OF
The Scots Armie.

New-castle, 27 *June* 1646.

EDINBURGH:
Printed by Evan Tyler, *Printer to the*
Kings most Excellent Majestie,
1646.

Courtesy Newcastle Society of Antiquaries

EDINBURGH, the ninteenth day of *July*, 1681.

FOraſmuch as it hath pleaſed the Kings moſt Excellent Majeſty, to call a Parliament of this Kingdom to meet at *Edinburgh*, upon *Thurſday* the twenty eighth of *July* inſtant; His *Royal Highneſs*, His *Majeſties* high Commiſſioner, and Lords of Privy Council, Do hereby Command and Require, the Lyon King at Arms, and his Brethren Heraulds, Macers, and Purſevants, to paſs to the Mercat Croſs of *Edinburgh*, and there, by open Proclamation, to Warn all the Lords Spiritual and Temporal, Officers of State, Commiſsioners of Shires and Burrows, to conveen at *Halyrudhouſe*, the ſaid twenty eighth of *July*, by nine of the Clock in the forenoon; the Lords furniſhed with their Gowns, Robes, Horſes, and Footmantles; and the Commiſsioners of Shires and Burrows, with their Horſes and Footmantles, to attend His *Royal Highneſs*, His *Majeſties* high Commiſsioner, on Horſe-back, in his going to the Parliament-Houſe, and in his return. And to the end, it may be known who are to be Admitted as Commiſsioners for Shires and Burrows, that all Perſons having Commiſsion from any Shire or Burgh, deliver in their Commiſsions to the Clerk of Regiſter, upon the twenty fifth inſtant, in the forenoon, in the Parliament-Houſe, that accordingly they may be Admitted as Commiſsioners to this Parliament.

WIL. PATERSON, Cl. *S*ti. Concilij.

Printed by the Relict of *Andrew Anderſon*, Printreſs to His moſt Sacred *Majeſty*, *Anno DOM.* 1681.

IX

LONDON AREA LOCATIONS

In his excellent series of articles on "English Provincial Presses' in Bibliographica from 1891, W. H. Allnutt includes as follows pp 289/290:-

XIX.—FINSBURY 1646-59

Dangerous errors in Several late printed Bibles to the great scandal, and corruption of sound and true Religion. Discovered by William Kilburne Gent ... *Printed at Finsbury Anno* 1659. 4[to]. [A copy in the Bodleian Library.]

The author of this tract complains that during the Civil War, through the absence of the King's Printers, and cessation of Bible-Printing at London, many erroneous English Bibles were printed in, and imported from, Holland, which by the late Assembly of Divines, were reported to Parliament in 1643, and condemned to the fire. The said Assembly, by direction of Parliament, then proposed the Bible printing to several Stationers of London, who refused; upon which it was commended to *Mr. William Bentley*, Printer in *Finsbury*, and his partners, who printed several Bibles in 8° and 12° in the years 1646, 1648, 1651, etc. The author goes on to complain that Bentley has been unjustly obstructed by *Mr. Hills* and *Mr. Field*, who have endeavoured to monopolise the sole printing of Bibles since the latter end of 1655.

Copies of Bentley's Bibles of 1646 and 1648 are in the Bodleian Library; but the imprint used is *London, Printed by William Bentley*. In an Ordinance of Parliament, however, made in 1649 special provision is made for the press in the city of York, and for the 'printing-press now used in Finsbury' in addition to those of London and the two Universities.

This historical linkage of Finsbury and York is of course of special interest to the present writer. This Act of Parliament: 'for better regulating the Press' was passed in September of 1649, four months after the Commonwealth had been declared in the months following the King's execution.

This Finsbury inclusion by Allnutt in his "Provincial Press" series raises the question of what were the boundaries of London in the 16th and 17th centuries, for today Finsbury is London N4 and N22, with Finsbury Park but a few high-speed minutes northwards from Kings Cross station, and within sight of Arsenal Football Stadium.

'London's boundaries through the centuries' is an intriguing and important subject which I will presently leave unresolved, except to note that printing commenced I believe:

(a) In SOUTHWARK, in 1514:
Cotton records from this "borough in Surry, on the Thames, opposite to the metropolis" [sic] that Peter Treveris established a press there in 1514 or 1516. Also from about 1536-8 James Nicholson printed several books including an English Bible and some New Testaments.

(b) In GREENWICH, in 1554(?):
The Rev. Henry Cotton in his Typographical Gazetteer Attempted of 1825 notes on page 63, "a small tract published during the reign of Queen Mary entitled A faythfull Admonycion..... Its colophon states "Imprynted at Grenewych by Comrade Freeman, in the month of May 1554". A copy is in the Bodleian Library and "the types, spelling etc. all prove the volume to have been executed in Switzerland or the Low Countries."

(c) Also in LAMBETH in 1572
W. H. Allnutt writes the following:-

'I have within my house in wagis, drawers and cutters, paynters, lymmers, wryters, and boke-bynders.' So says Archbishop Parker, writing to Lord Treasurer Burghley, under date May xiiii, 1573, and sending him three books, one of which was his own private production,

> De antiquitate Britannicæ Ecclesiæ & Priuilegiis Ecclesiæ Cantuariensis, cum Archiepiscopis eiusdem 70. An. Dom. 1572. Folio.

This title is within an elaborately emblematic, but inappropriate woodcut compartment, representing Ptolomeus, Marinus, Aratus, Strabo, Hipparchus, Polibius, Geometria, Astronomia, Arithmetica, Musica, and a fat Mercurius fully clothed in doublet and jerkin, seated on clouds with the caduceus in his right hand. This same emblematic cut was much more appropriately used some years before for the title-page of William Cuningham's *Cosmographical Glasse*, 'Excussum Londini in officina Ioan. Daij Typographi. Anno 1559.'

That the Archbishop's book was printed by John Day is generally acknowledged; but that the printing was performed in Lambeth Palace has been disputed. The Bodleian Library possesses two copies, in one of which is the following inscription: '5to Januarij 1593. Hunc librum dono dedit Richardo Cosin Johannes Parker armiger filius primogenitus Matthei Parker nuper Cantuariensis Archiepi cuius auspiciis et sumptibus liber iste et collectus et impressus est proprijs in ædibus Lamethæ positis.' It seems, therefore, that the book was really printed at Lambeth, and that Day supplied the workmen, printing-press, and materials, and probably exercised a general superintendence. Strype, in his *Life of Parker*, iii. 344, gives a list of the Archbishop's debts and funeral charges, among which occurs, 'To Richard Ing, printer, in full 4. 0. 0.'

Only a very limited number of the book under discussion were printed, and the majority of these were still undistributed when the Archbishop died. Martin, in his *Catalogue of Privately Printed Books*. 1854, says: 'It has been observed that no two copies of this work have been found alike; and I can bear witness to the truth of this assertion in those I have collated.' Martin devotes fourteen pages to his description of the book and the variations of the twenty-one copies of which he gives the whereabouts.

PRINTING IN CHESTER: 1688?

My good friend Derek Nuttall, who gained his Doctorate of Reading University in 1985, with his epic, profusely illustrated thesis of more than 485 pages: English printers 1600-1700 and their supra-text roman and italic types (of which I have a treasured autographed copy in my personal possession), had earlier in 1969, published a well-researched History of Printing in Chester from 1688 to 1965. With appreciation I reproduce on the following page from that book the 1688 Chester Title page The Academy of Armory.... by "Randle Holme, of the City of Chester, Gentleman Sewer in Extraordinary to his late Majesty King Charles 2". Derek records three extant copies of it: in Chester Public Library and in the University libraries of Aberystwyth and of Keele.

Academy of Armory was a substantial book of more than 500 pages, of a large page size of 8-5/8" x 13½". An unusual feature is that it has two title-pages, the one with letterpress type as illustrated, and also a copper-plate version which specifically states "Printed att Chester by the Author". Nuttall also quotes supporting contemporary letters, so that there is very strong supposition for this 1688 Chester press, which produced this one work of substance. The probably printer was "T. Tillier, Typog."

After that there are no indications of Chester printing until early in the following century, when from 1702/3 the minute book of the Chester Stationers Company record payments made to "Travelling Printers" on four separate occasions. (For full and interesting details, see Derek Nuttall's Chester book especially pages 4 to 15).

THE

ACADEMY OF ARMORY,

OR,

A STOREHOUSE

OF

ARMORY

AND

BLAZON.

CONTAINING

The ſeveral variety of Created Beings, and how born in Coats of Arms; both Foreign and Domeſtick.

WITH

The Inſtruments uſed in all Trades and Sciences, together with their Terms of Art.

ALSO

The Etymologies, Definitions, and Hiſtorical Obſervations on the ſame, Explicated and Explained according to our Modern Language.

Very uſeful for all Gentlemen, Scholars, Divines, and all ſuch as deſire any Knowledge in Arts and Sciences.

Every Man ſhall Camp by his Standard, and under the Enſign of his Fathers Houſe, Numb. 2. 2.
Put on the whole Armour of God, that you may be able to ſtand againſt the Aſſaults of the Devil; above all take the Shield of Faith. Epheſ. 6. 11. 16.

By *Randle Holme*, of the City of *Cheſter*, Gentleman Sewer in Extraordinary to his late Majeſty King *Charles* 2. And ſometimes Deputy for the Kings of Arms.

CHESTER,
Printed for the Author, MDCLXXXVIII.

Courtesy <u>A History of Printing in Chester</u>
by Derek Nuttall (1969)

XI

FURTHER PRINTING IN EXETER: 1688 - 1689

The first printing in Exeter in 1645-1646 by King Charles' travelling printers with their two presses is recorded, with illustration of all known extant imprints, in the second of my King's Printer 'green-back' studies. Additionally I illustrate on pages 126 and 127, facsimiles of two 1648 productions, the second of which states "Printed at Exeter".

Then at the time of Dutch William of Orange's landing in Tor Bay in the autumn of 1688, an Exeter printer appeared from nowhere, to print four small but important titles for dissemination in the cause of the Protestant invader.

At that time printing in Exeter or anywhere else in the English provinces, save Cambridge, Oxford and York, was not permitted unless a printer was already a member of the London Stationers Company, and there is no such evidence for Exeter in 1688. The two most likely possibilities are either that William of Orange brought his printer with him (just as his wife's grandfather Charles I had found so useful in 1639 and again from 1642-1646 and just as William himself was to have with his army at the Battle of the Boyne in Ireland only two years later, in 1690). Alternatively when Francis Egglesfield the London bookseller bought the two Exeter presses from the disgraced and defeated Royalist printer John Bill (see page 69 of my 'green-back' study) in the spring of 1646, at least one of the presses may have lingered secretly and semi-dormant in the Exeter region, until times were more propitious, as they so dramatically did in the southwest of England in the autumn of 1688, just three hundred years ago.

W. H. Allnutt in his "English Provincial Presses" series in Bibliographica, records as follows:-

On the occasion of the Revolution we again find a press at work in Exeter, from which issued :—

1. The speech of the Prince of Orange, to some principle gentlemen of Somersetshire and Dorsetshire, on their coming to joyn his Highness at Exeter the 15th of Nov. 1688. *Exeter, Printed by J. B.* 1688. Broadside.
2. A Form of Prayer, &c. [For the Prince and Princess of Orange.] Translated from the Dutch. *Exeter, Printed by J. B.* 1688. 4to.
3. The General Association of the Gentlemen of Devon, to his Highness the Prince of Orange. *Exon: Printed in the Year,* 1689. Broadside.
4. The Prince of Oranges Speech to the Lords, &c. concerned with him. *Printed at Exeter,* MDCLXXXIX. 1 leaf, folio.

A copy of each is in the Bodleian Library.

Derek Nuttall, in his Reading University Theses (page 124) also records a copy in Guildhall Library London (Broadside 7.75) with the same title as Allnutt's No. 3, but with the year date of 1688.

Thus far no-one has identified this "deus-ex-machina" appearing 1688 Exeter printer with his stated initials: J.B. , despite there being on-the-spot eminent printing historians in Exeter named Maxsted, Stirling and colleagues. Who I wonder will be the first to be the 'J.B. code-breaker'?

XII

IN CONCLUSION

In addition to several 'London Area Locations' briefly mentioned in Chapter IX, other pre-1695 British printing locations include:-

(a) MANCHESTER: 1589 and 1664? :-

As has been described in the Marprelate section, Manchester was the scene of the August 1589 capture of the secret press while it was type-setting (but before printing) More Worke for Cooper. Later W. H. Allnutt in his Notes on Printers and Printing (1878) lists 1664 as possibly the date of Manchester's first printing. This work was titled A Guide to Heaven from the Word and Allnutt quotes his source as Local Gleanings relating to Lancashire and Cheshire, edited by J. P. Earwaker, Manchester in 2 volumes 1875 and 1878. Allnutt also usefully lists the first printing date for very many places in England and Wales during the eighteenth century.

(b) COLCHESTER: 1648:-

Clough records one quarto dated 1648 "A choak-peare for the Parliament". Printed at Colchester (Wing C.3921).

(c) ROCHESTER: 1648? and 1688:-

W. H. Allnutt in his "English Provincial Presses" series of articles (p 279) expresses doubt about a 1648 8-pager, The Kentish Fayre "printed at Rochester and are to be sold, to all those that dare buy them". Allnutt (p 279) quotes from it

> "Those cursed Traytors that would kill their King,
> Unto a hanging we will quickly bring"

Then Clough lists two from 1688 and 1689 both titled: "James II, King of England, His Majesties reasons for withdrawing himself from Rochester." (Wing J376 and J377).

EPHEMERA

In my 'green-back' series I have recorded many single-sheet pieces of printing ranging from proclamations to pre-execution verses to early printed Indulgences. Other interesting but elusive species of ephemera undoubtedly include Playing-Cards and also printed Book-seller labels. As to Wall-papers, I have in my first 'green-back' Printers Dozen (pages 50-52) described and also illustrated what may have been Britain's earliest extant printed wall or beam coverings, which may be seen in the Library of Christ's College Cambridge carefully preserved between glass. These may date from 1509 and were possibly printed, from intriguing pictorial evidence, by Hugo Goes, once of York and of Beverley.

Printing then, as now, was not only concerned with books.

SUPPLY INDUSTRIES

Printing in these divers locations was clearly dependent on access to supplies of paper, ink, types and presses. A largely-new and separate subject would be to take each of the smaller printing localities described in my 'green-back' series: to seek out from my facsimile data the originals of the books, tracts and broadsheets there printed: to study the paper water-marks (which I myself have only found time to check very spasmodically): and then to seek to link such paper water-marks with general and specific studies of the contemporary paper industry. One aim would be to discover to what extent local British printing, in these first two centuries after Caxton, was (or was not) dependent on localised supplies of paper (a) in times of civil war and (b) in times of peace. My great hope is that someone will 'pick up this torch', for there is so much more needing to be discovered about the early spreading of printing through Britain. (I myself am in my early seventies and about to commemorate fifty active years in printing administration. Yet it was only the age of 79 that caused Falconer Madan to end his classic Oxford Books series at 1680, rather than at 1700 as he had originally hoped!).

XIII

PRINT OUTPUT CHARTS

XIII

OUTPUT CHARTS

Firstly as regards the period of my Printers Dozen 'green-back', i.e. up to the Stationers Company charter year of 1557, W. H. Allnutt comments, at the end of his first "English Provincial Presses" article in Bibliographica in 1891, that "just over one hundred books" had been printed on English provincial presses prior to this Stationers Company restrictive charter.

Covering the whole period of my printing history studies (to 1695 and indeed beyond), Ian Maxted has prepared the following graph which I reproduce here with great appreciation to him. This records the total extant output of British imprints from 1500, showing a tremendous surge of titles around the beginning of England's civil war, in the early 1640s; another surge, but not so large at the 1660 restoration of the monarchy; thence climbing to new heights towards the end of that century.

The print output graphs which then follow: separately for Cambridge and Oxford; for Ireland and Scotland; and for York are necessarily on a reduced scaling of output. Thus a quick flip through the series of graphs will demonstrate that the print output of LONDON was throughout this period dominant and massive in comparison with the scattered series of other British printing locations studied in my 'green-backs'.

The next print-output-chart, that for CAMBRIDGE, is based totally on the year-by-year listing shown in E. A. Clough's A Short-Title Catalogue Arranged Geographically...... This having been published in 1969, was necessarily taken from the first editions of STC and of 'Wing'. The results, as shown in my series of graphs of the individual locations, are therefore clearly incomplete.

However, they do I believe high-light different scales of print output, at different places and at differing periods of national and of international history. It is hoped that this present partial exercise will encourage others, with computer assistance, to make further up-to-date, geographic year-by-year, print-output analyses based of course on the second editions of STC and of 'Wing'.

Clough in his Preface clearly states that for Scotland, for Ireland and for Oxford, his own 1969 geographic listing merely supplements the already existing title-lists (a) by Harry Aldis for Scotland; (b) by Mclintock Dix for Dublin; and (c) by Falconer Madan in his 3-volumed Oxford Books.

For these three locations therefore, I have counted and added each of their year-by-year totals to the equivalent Clough totals for the purposes of these individual-location graphs. For Dublin, I have also added the supplementary titles listed in Dix' second volume from page 325. For Scotland I have added together the titles produced in Scots places other than Edinburgh, which city has the substantially largest output of the Scots locations.

If there are in fact duplications between Clough and these other three listings, I have not eliminated them. Furthermore the listings include some titles "printed for" rather than "printed at" which again would need computer assistance to be sure of confining the numbers to the particular location's print output year by year. Again works which are in fact un-dated have been totalled into the year in which they have been listed by the above-mentioned four authorities, which introduces a further subjective element.

Another qualification is that anyway for Scotland, a series of Journal printings has only been counted as one. (See for example Aldis' entries 1650, 1651 and 1652 for the year 1660).

Throughout it is important to remember that all such listings, however up-to-date and meticulous they may become, still only record the EXTANT titles which have chanced to survive burnings and civil-wars and lesser hazards to the present-day. The most reliable indicator of this known to me is the Contemporary List of the King's travelling printer at York in 1642 and at Shrewsbury in 1642-1643. In great appreciation to the National Library of Scotland, I reproduced this unique list complete in my first King's Printer volume (pages 32 to 44). The contemporary list gives short titles, dates and sizes of work for 166 named titles; and my volume describes and illustrates my extensive search for copies of these which are still extant. From this secure base of contemporary listing comes the dramatic (and still surprisingly little noted result) that (a) in the months before the Civil War fighting started (the first part of 1642) about seven-tenths of York's actually produced output survives; whereas (b) in the ten months after the outbreak of fighting (October 1642 to August 1643) only about one-fifth of Shrewsbury's actually-produced 96 titles have survived. (see my summary page 54 in that volume).

This basis of secure and definite knowledge of substantial total loss of printed titles is certain to be greater for the more ephemeral single-sheets and smaller booklets than for larger works. On this known basis, each reader is invited mentally to 'pencil in' their own much higher line on this series of graphs: say actually higher by one-third or even by one-half (though probably not by the actual four-fifths loss rate of the Shrewsbury 1642-1643 output).

Yet another factor calling for comment is that with all of these listings, one single-sheet proclamation and one Bible - each count as one title on these graphs. A count of the sheet-by-sheet print output would be the scientific answer, plus an estimate of quantities produced whether of proclamations, tracts or Bible. I will very happily leave someone else's computer to tackle these elusive and sizable tasks!

The following CAMBRIDGE graph is fairly steady and noticably lacking in printing surges relative to other graphs. By contrast OXFORD's graph records a tremendous output surge to 293 titles in 1642 when Charles I established his military headquarters there that autumn. Madan records no Oxford printing at all between October 1648 and May 1649. A further smaller Oxford surge occurred in the Restoration year of 1660, but there was no marked increase of title output around the 1688 year of Dutch William's invasion.

For the early Civil War period, attempting to assess Oxford's year by year output is additionally hazardous because there were a large number of counterfeit Oxford imprints. Royalist sympathisers in London evidently found it less risky to take a single easily-concealed copy of an Oxford-printed tract into London and then to arrange for it to be secretly printed in the metropolis - with a false Oxford imprint. Falconer Madan (p x) writes; in 1642, "Out of 191 [sic] Oxford imprints, no less than 58 are London counterfeits; in 1643, 41 out of 238; and in 1644, 24 out of 145. At one period (March 25-April 17, 1644) there are as many false imprints as there are genuine."

Also as between centuries, Madan (p xii) makes the interesting broad observation that the Oxford printers' "entire output of the eighteenth century is actually less than that of the seventeenth".

The DUBLIN print output graph shows surges in 1641, in 1661, in 1685 (Accession of James II); also around 1690 (Battle of the Boyne) and markedly in 1695, Dublin's first year above the one hundred title level.

SCOTLAND nearly reached this one hundred title level in 1648, and passed it in 1660: similarly in 1680, 1681 and 1683. James II's Accession year of 1685 produced 184 Scots extant titles; with the highest annual output that century, of 271 titles in the 1689 Accession year of William and Mary; and with outputs continuing usually well above the one-hundred-title level year by year until for 1700 the total became 232 extant titles.

Meantime YORK had its initial royalist printing surge in 1642 with 78 actual titles by King Charles' travelling printer, plus I reckon 25 titles produced by the other incoming royalist Stephen Bulkley, in the second part of that same year, so that in 1642 York's actual print output also passed the one-hundred-title-level, a York print record which was to last for a very long span of time.

For the year 1643, Stephen Bulkley (continuing in York) produced 25 printing titles; then in the following year came the royalist defeat at Marston Moor and the capitulation of York. After this York's print output continued with the Parliamentarian arrival of Thomas Broad and so proceeded at a modest level and probably continuously, though with some gaps in the annual listings of extant titles, through the years into the eighteenth century.

All in all therefore, will readers please take these following graphs, not just with the proverbial pinch of salt; but rather with a whole pillar of salt (such as was the fate of Lot's wife).*

So now, E. & O. E., please turn over and ponder upon the reasons for the often low-levels of non-London print-output, offset by occasional dramatic print-output surges in different localities, some at similar and some at differing timings.

* "But Lot's wife [despite the angels' warning not to do so] looked back [upon the Lord raining fire and brimstone from the skies upon the wicked cities of Sodom and Gomorrah] and she turned into a pillar of salt" - Genesis Chapter 19 Verse 26.

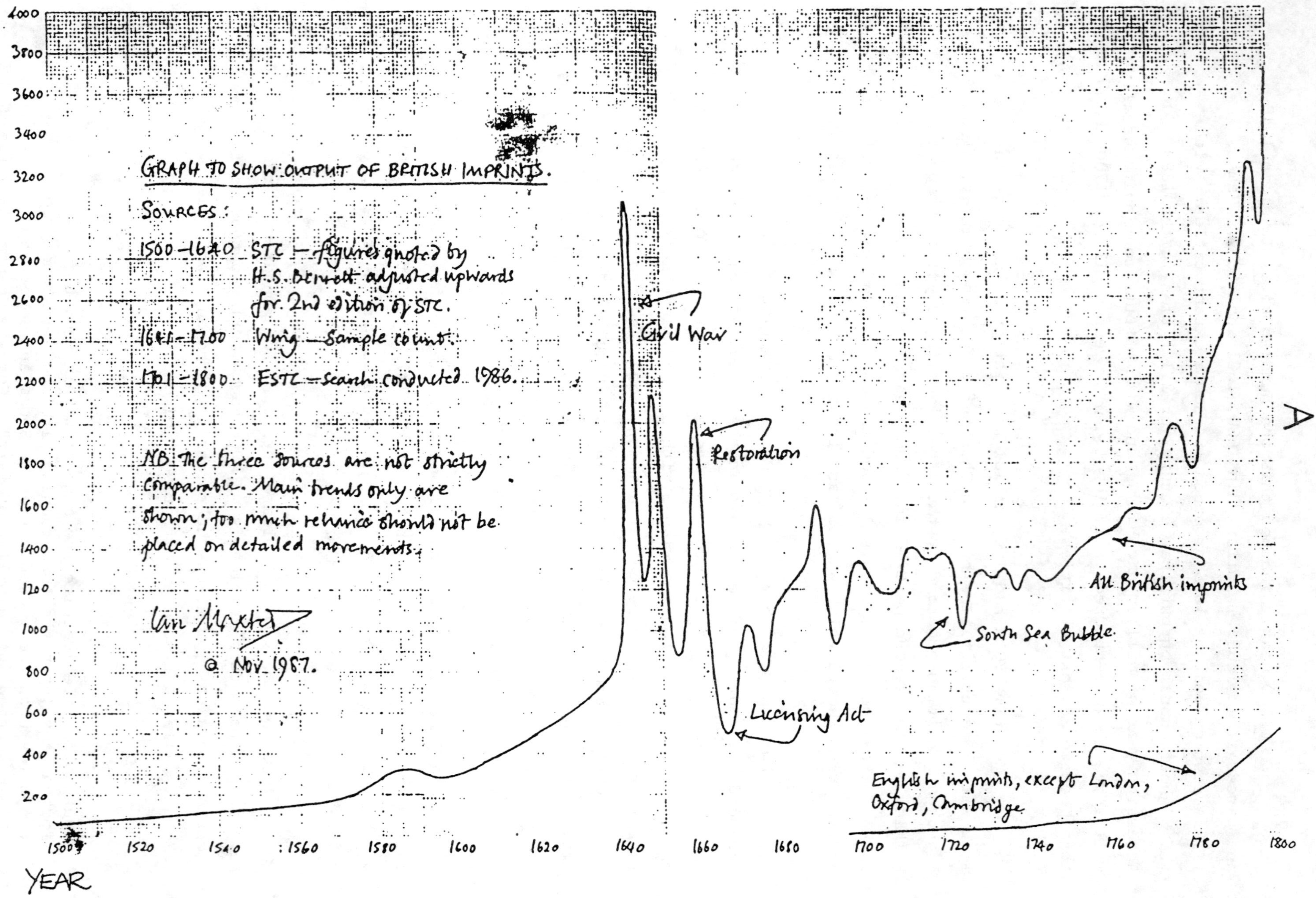
NUMBER OF TITLES
4000
3800
3600
3400
3200
3000
2800
2600
2400
2200
2000
1800
1600
1400
1200
1000
800
600
400
200
GRAPH TO SHOW OUTPUT OF BRITISH IMPRINTS.
SOURCES:
1500–1640 STC — figures quoted by H.S. Bennett adjusted upwards for 2nd edition of STC.
1641–1700 Wing — sample count.
1701–1800 ESTC — search conducted 1986.
NB The three sources are not strictly comparable. Main trends only are shown; too much reliance should not be placed on detailed movements.
© Nov 1987.
Civil War
Restoration
Licensing Act
South Sea Bubble
All British imprints
English imprints, except London, Oxford, Cambridge
A
1500
1520
1540
1560
1580
1600
1620
1640
1660
1680
1700
1720
1740
1760
1780
1800
YEAR

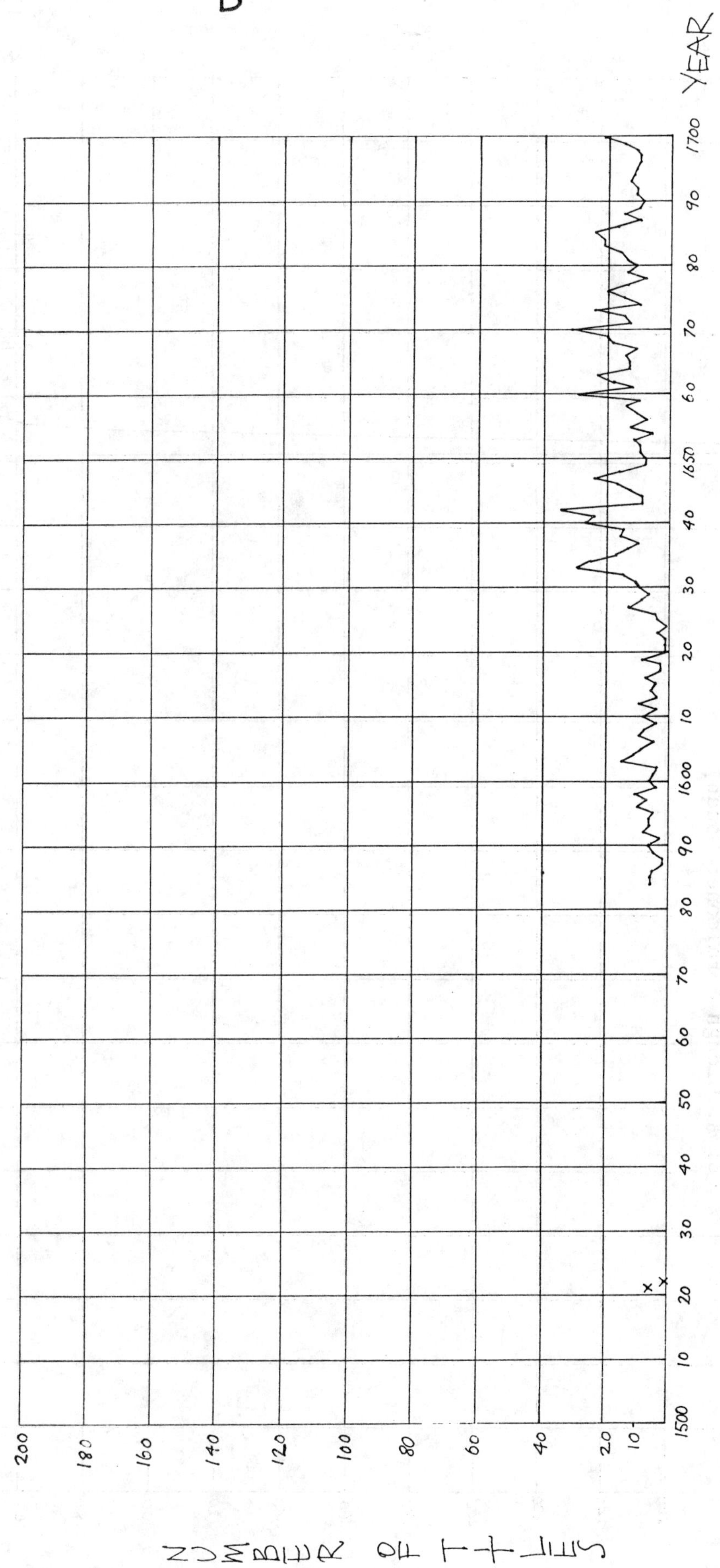

Output of Extant CAMBRIDGE Printed Titles
(E. A. Clough)

YEAR

1500 10 20 30 40 50 60 70 80 90 1600 10 20 30 40 1650 60 70 80 90 1700

NUMBER OF TITLES

200 180 160 140 120 100 80 60 40 20 10

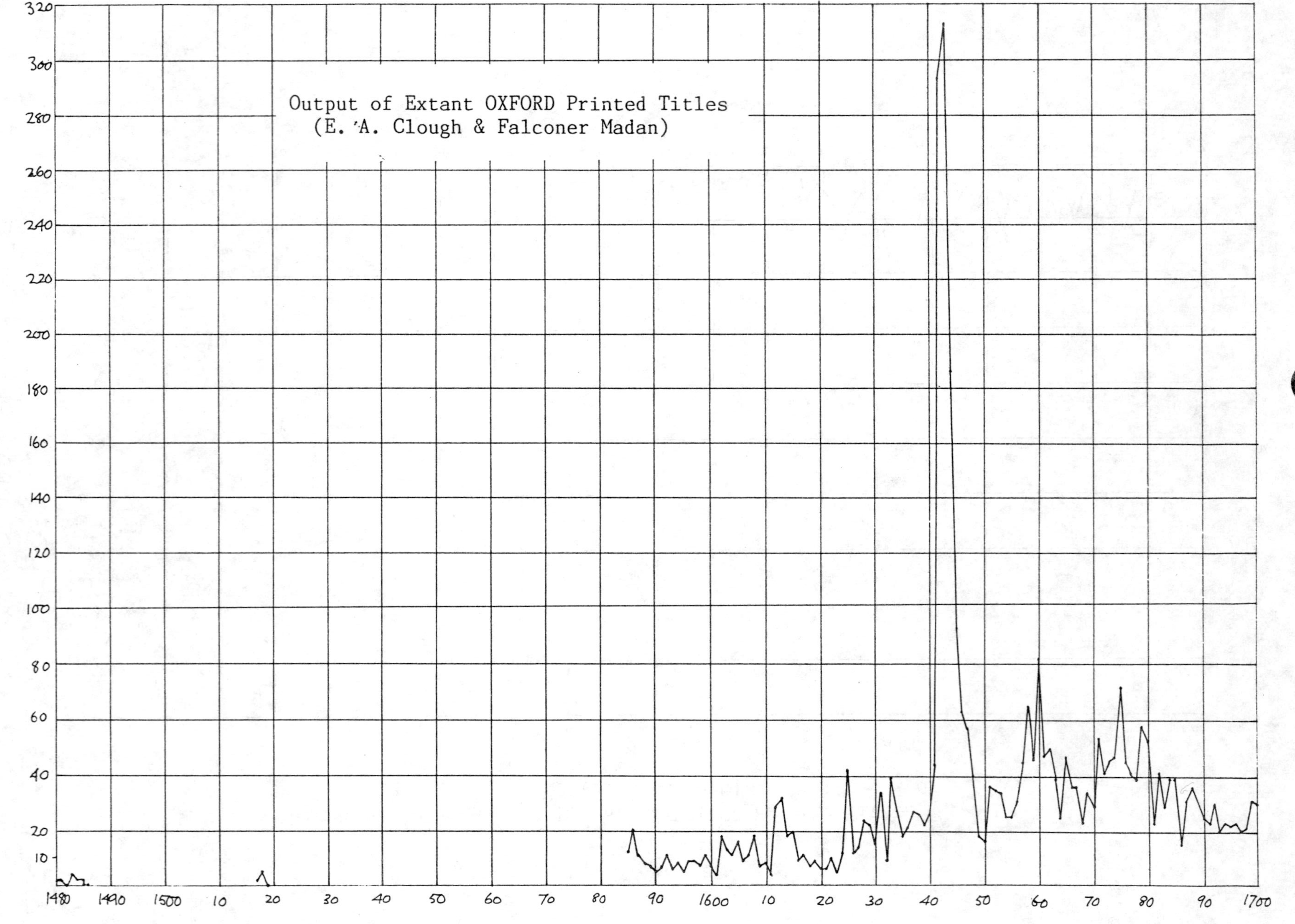
Output of Extant OXFORD Printed Titles
(E. A. Clough & Falconer Madan)
C
NUMBER OF TITLES
320
300
280
260
240
220
200
180
160
140
120
100
80
60
40
20
10
YEAR
1480
1490
1500
10
20
30
40
50
60
70
80
90
1600
10
20
30
40
50
60
70
80
90
1700

D

Output of Extant DUBLIN Printed Titles
(E. A. Clough & McClintock Dix)

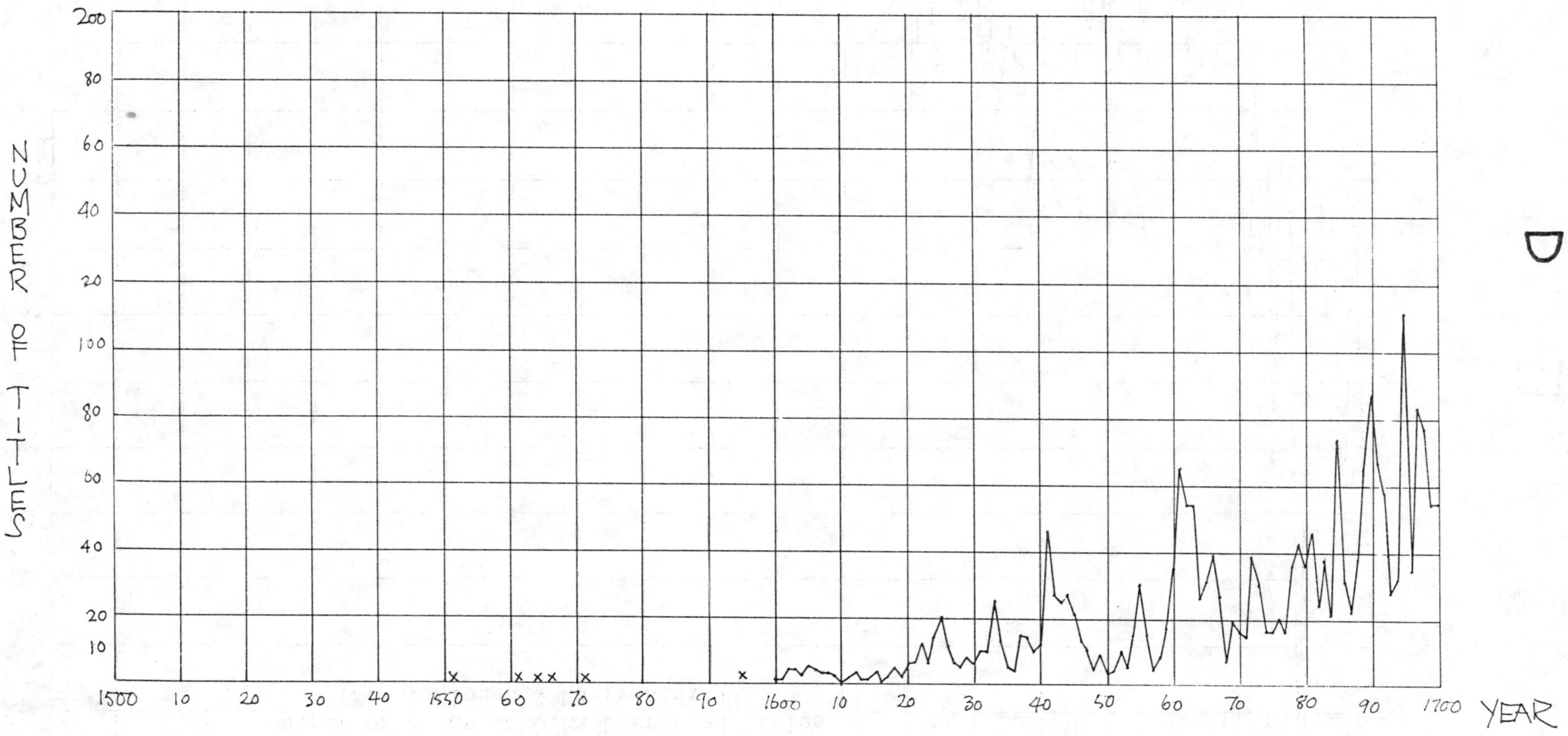

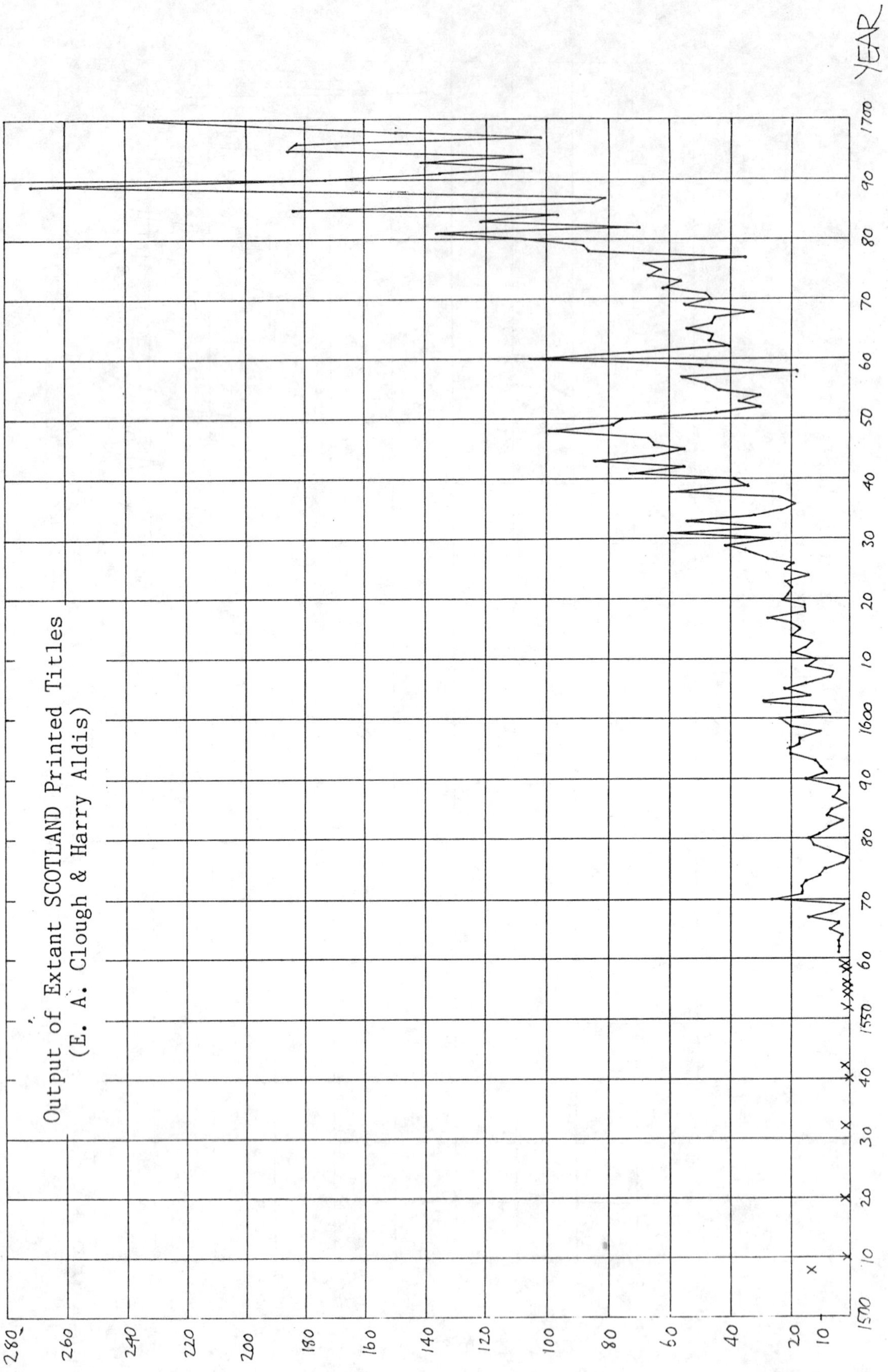

Output of Extant SCOTLAND Printed Titles
(E. A. Clough & Harry Aldis)

YEAR

1500 10 20 30 40 1550 60 70 80 90 1600 10 20 30 40 50 60 70 80 90 1700

NUMBER OF TITLES

280 260 240 220 200 180 160 140 120 100 80 60 40 20 10

F

Output of Known YORK Printed Titles
(E. A. Clough & Contemporary 1642 Catalogue)

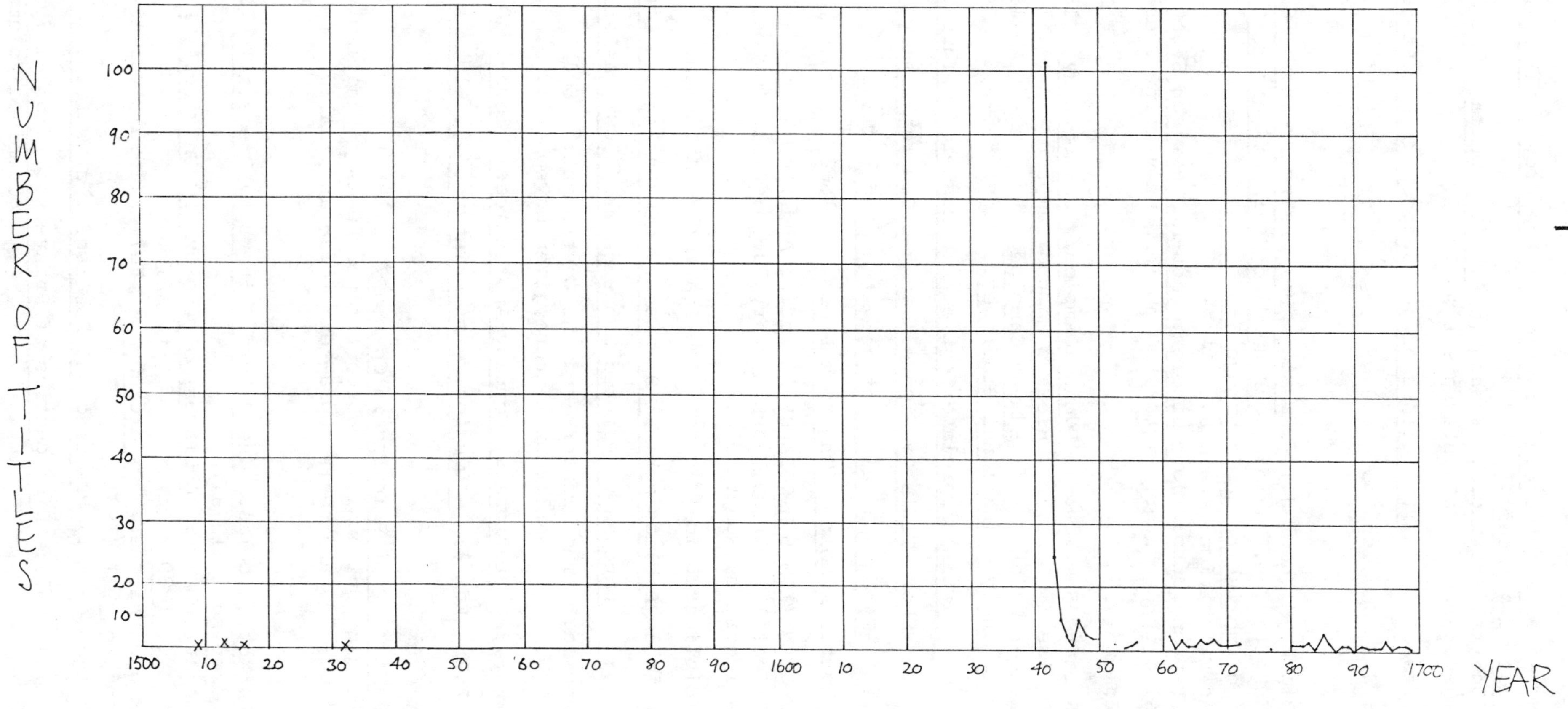

NUMBER OF TITLES

100
90
80
70
60
50
40
30
20
10

1500 10 20 30 40 50 60 70 80 90 1600 10 20 30 40 50 60 70 80 90 1700 YEAR

BIBLIOGRAPHY

ALLNUTT, W. H.

(a) Notes on Printers and Printing in the Provincial Towns of England & Wales (1879)
(b) "English Provincial Presses" in Bibliographica (Several parts at intervals from 1891).

ARBER, Professor: Transcripts of the Stationers Registers.

BENNETT, H.S.: English Books and Readers Vol II (1588-1603); Vol III (1603-1640)

CAMBRIDGE:

(a) Black, Michael H.: Cambridge University Press: 1584-1984 (1984)
(b) McKitterick, David: Four Hundred Years of University Printing and Publishing in Cambridge 1584-1984: An Exhibition Catalogue. (1984)
(c) Trepton, Otto: John Siberch: Johann Lair von Sieburg (1970)
(d) Also three "Short History" Cambridge University Press booklets.

CAMPION, Edmund:

(a) Richard Simpson: Edmund Campion: a Biography (1867)
(b) Louise I Guiney: Blessed Edmund Campion (1908)
(c) Evelyn Waugh: Edmund Campion (1937)
(d) Dictionary of National Biography: Pages 850-854.

CARLSON, L.H: Martin Marprelate Gentleman: Master Job Throkmorton (1981)

CLOUGH, E.A.: Short-Title Catalogue Arranged Geographically (1969)

COTTON, H.: Typographical Gazetteer Attempted (1825)

DAVIES, Robert: Memoir of The York Press (1868)

DICTIONARY OF NATIONAL BIOGRAPHY including Anderton, James ('John Brereley Priest'); Norton, John; Penry, John; Throkmorton, Job; Waldegrave, Robert etc.

EDINBURGH Bibliographical Society Publications: Articles on Scots Printers.

GILLOW, Joseph: Bibliographical Dictionary of the English Catholics.

HAWKES, A.J.: Lancashire Printed Books before 1800: a Bibliography (1925)

LAMBERT, Sheila: "The Printers and the Government: 1604 to 1637" - Script of lecture given November 1986.

McKENZIE, D.F.: Stationers' Company Apprentices: 1605-1640 (University of Virginia, 1961)

McKERROW, R.B. (Edit): A Dictionary of Printers...... 1557 to 1640

NUTTALL, Derek

(a) A History of Printing in Chester (1969)
(b) Reading University 1985 Doctorate Thesis: "English Printers 1600-1700 and their supra-text roman and italic types"

OXFORD

(a) Barker, Nicolas: The Oxford University Press & The Spread of Learning (1978)
(b) Carter, Harry: A History of the Oxford University Press... to.... 1780 (1975)
(c) Madan, Falconer: Oxford Books: Vol I (1895), Vol II (1912) and Vol III (1931)
(d) Morgan, Paul: Printing and Publishing at Oxford.... (1978)

PLOMER, H.R. (et al): Dictionaries of Printers.... 1557-1775 (1977 Combined Volume)

SCOTLAND: (a) Dickson, Robert and Edmond, John Ph.: Annals of Scottish Printing: 1507 to 1610 (1890 and 1975 reprint)

(b) Aldis, Harry G.: A List of Books Printed in Scotland Before 1701 (1904 & 1970 reprint)

SESSIONS: See Back Cover

STATIONERS' COMPANY

(a) Blagden, Cyprion: The Stationers' Company, a History 1403-1959 (Allen & Unwin, 1960)

(b) Pollard, Graham "The Company of Stationers before 1557" in The Library 4th series xviii, 1938

(c) Anon: Quater-Centenary: 1557 to 1957 (Stationers Hall 1957)

WICKLEN Stanley I: Princes and Printers (Cardiff: 1969).

ACKNOWLEDGMENTS

I continue to be most grateful to many national and international advisers, some mentioned in this text and others in previous acknowledgments in this series.

Expert typing help has once again been rendered by Mrs Margaret Atkinson (my Ebor Press Secretary from 1964 to 1987). My thanks also go to J Bryan Blackwell for skills with maps, graphs and facsimiles (including the important mm/inch scaling); and most of all to Margot my wife for her support, even to the extent, during one final-final weekend, of not being able to eat at her own history-filled dining table! We will long remember our Easter 1988 visits to Eton College and Stonor Park for example: (though we have yet to look for the seaside printing cave near Llandudno: see page73)!